The Bio-Origins Series

View of the earth as photographed from the Apollo 17 spacecraft, the final lunar landing
mission in NASA's Apollo program.

GENESIS ON PLANET EARTH

The Search for Life's Beginning

SECOND EDITION

WILLIAM DAY

Foreword by Lynn Margulis

Yale University Press
New Haven and London

Published with assistance from the foundation established in memory of Amasa Stone Mather of the Class of 1907, Yale College.

Designed by Nancy Ovedovitz and set in VIP Times Roman type. Printed in the United States of America by Murray Printing Company, Westford, Massachusetts.

The paper in this book meets the guidelines for permanence and durability of the Committee on Production Guidelines for Book Longevity of the Council on Library Resources.

Library of Congress Cataloging in Publication Data

Day, William, 1928–
  Genesis on planet Earth.
  Bibliography: p.
  Includes index.
  1. Life—Origin. 2. Paleontology. I. Title.
QH325.D35 1984   577   83–21900
ISBN 0–300–02954–3
ISBN 0–300–03202–1 (pbk.)

10 9 8 7 6 5 4 3 2 1

Nature set herself the task to catch in flight the light streaming towards the earth, and to store this, the most evasive of all forces, by converting it into an immobile form. To achieve this, she has covered the earth's crust with organisms which while living take up the sunlight and use its force to add continuously to a sum of chemical difference.

Julius Robert von Mayer
Die organische Bewegung in ihrem
Zusammenhange mit Dem Stoffwechsel, 1845

Contents

Figures

Frontispiece: View of the earth as photographed from the Apollo 17 spacecraft during the final lunar landing mission in NASA's Apollo program (courtesy of NASA).

Tables

Foreword

In midsemester, 1979, amidst the bustle of a huge teaching commitment, I received by book post the first edition of *Genesis on Planet Earth*. William Day, the author, and House of Talos, the publisher, were entirely unknown to me. A quick scanning of the sans serif type revealed that the text was interspersed with formulae of photosynthesis and respiration written with a manual typewriter. Statements such as "Species have followed species, rung after rung, in a continuous climb of the ladder called evolution" and "Man, himself, will not change appreciably or advance beyond his technological society, simply because technology has replaced biological evolution" seemed to bounce off the page.

Day seemed to be perpetuating the usual evolutionary misunderstandings. Much time in my two evolution courses is spent in countering just such loose statements as these. We reject the nineteenth-century "typological view" so consistent with the Aristotelian "ladder of life" concept. In contrast to the notion of progressive evolution of types toward superior beings we develop ideas of the origin and changes of genes and frequency distributions of characters in natural populations. To deny that people are still evolving is silly. We show quantitatively how natural selection is acting on various human populations today.

Moreover, every week I must battle an influx of pseudoevolutionary, often religious, nonsense on would-be evolutionary themes: phone calls, letters, articles, even books. Not surprisingly, then, after one cursory glance I set down Day's version of Genesis in the "philosophically reject" pile. With its bright red ribbon sewn in, it struck me more as a hymnal than a text of natural science.

I was disappointed, though, because the need for a comprehensive and comprehensible scientific text on the genesis of life is enormous.

Mitchell Rambler, my teaching fellow in Environmental Evolution at the time, and now head of the Exobiology program at the NASA Life Sciences Offices in Washington, picked the book off the reject pile and took it home with him. A week later he confronted me with a firm recommendation that we use Day's book as supplementary reading in the course for a unit of study on the origins and early evolution of life. "Read the parts on the origin of coupled nucleic acid and protein synthesis," Rambler urged enthusiastically. "Read the section on proteinoids." I took his advice and found myself reading the book from cover to cover. Amazed that I had so misjudged the book, I thanked Rambler profusely. We have used the book ever since.

One forgave Day's biological clichés and the flaws in his geological chapter. One ignored the homemade production of the volume, which was distributed from a private address in East Lansing, Michigan. One delighted in a clearly written and important book that tied together observations from astronomy, geology, microbiology, and, especially, prebiotic chemistry. The history of the giant intellectual problem of the origin of life was dealt with gracefully and in a manner totally accessible to students.

Now, in 1984, the text has been extensively rewritten. The flaws have receded and the book's strengths have been accentuated. The production job is vastly improved. Retaining its original clarity and breadth, this second edition is a fully professional work.

Day understands thoroughly how the origin of life has been transformed from a source of philosophical speculation into a problem of experimental and observational science. Day assumes, correctly, that the origin of life problem is entirely susceptible to scientific solutions. He sees that one major issue, the origin of monomers, only slightly transcends the concerns of ordinary chemistry. The enormous progress already made on that issue is lucidly reviewed.

But the subject quickly grows complex and confounds. The problem of the emergence of life from organic chemical monomers and their macromolecules is far more elusive. The origin of metabolizing autopoietic systems from chemicals is the central issue here. It is identical to the problem of the appearance of the minimal procaryotic cell. From biochemistry and molecular biology, on the near edge, this problem reaches into the province of anaerobic microbiology at a distant border. In describing this chemical–biological frontier, the crux of the matter, Day excels. He distinguishes the relevant experiments; he does the biologist and chemist an enormous service as he critically evaluates the irrelevant. He points out the present limitations to the various components of our explanation of the appearance of the self-sustaining chemical system called life. Day deals with genetics and cell biology, pre-Phanerozoic geology, and human evolutionary biology with less ease. The salient fact is, however, that he writes about them clearly and in a straightforward and interesting way.

The second edition of Day's book stands alone as the book we sought. It is an original, comprehensive, and comprehensible account of the scientific Genesis: the great drama that occurred on the surface of the Earth some four billion years ago. Furthermore it is designed, from the beginning, for students. Students and teachers, researchers and witnesses of our space age world will welcome this text and congratulate the Yale University Press for making it available to a wider audience.

Boston University   Lynn Margulis
January 1984   Professor of Biology

Preface

Until thirty years ago any meaningful study of how life began on primordial earth seemed beyond the pale of scientific inquiry. Even the simplest microorganisms consisted of extremely large and complex molecules called proteins, much too intricate for precise analysis. The key to life's reproduction was in the genes, where the synthesis of proteins was hypothesized to take place on some type of template. But since the constituents of proteins, the amino acids, occurred in nature only as products of biological systems, how could the first cells have formed?

Then, all in the same year, 1953, Frederick Sanger announced that he and his coworkers had resolved the primary structure of a protein molecule, James Watson and Francis Crick unraveled the chemical basis of life and found it to be a double-helical nucleic acid, and Stanley Miller discovered how matter and energy could create the building blocks of life without a preexisting living cell.

Unleashed from the paralysis and spurred by the Space Age, research on the origin of life was launched by these three momentous scientific achievements into an era of discovery and revelation. Old impressions were rapidly supplanted by new concepts as experiments changed the image of primordial earth—and the new scenario gave clues to life's beginning.

Today, the manner in which life first appeared on earth can be scientifically reconstructed. It is a fascinating story, both in the search and in the conclusion. The question that faced scientists in the search was: At what level of material existence was matter able to bridge the great chasm that divides the biological from the inanimate world?

To find the source of life's origin we have to return to circumstances that no longer exist, to an earlier age when the conditions were different from those which exist today. And we have to return to a different dimension, to the realm of the minuscule that lies far below our sensory perception, to a level where life exists closer to the interface that divides the living from the nonliving. For it is at this level and with these conditions of an earlier time that we find the threshold that led out of the inanimate world into the world of the living.

Acknowledgments

Most of the background for this book came from the many stimulating discussions I had with William Stillwell, Frank Denes, and others while we were at the Institute for Molecular and Cellular Evolution in Miami. The invaluable critiques of William Stillwell made possible the first edition.

I am deeply indebted to Lynn Margulis for her support and suggestions for the revised edition. My gratitude goes also to Louis Brown for his corrections concerning energy and astronomy, to Ben Stoller for the exchange of ideas and material, to Laura Randall for her encouragement, and to the many readers of the first edition who sent heartening comments.

George Hildenbrandt, Dennis O'Callaghan, John Basford, James Corbin, and Ben and Sandy Van Osdol read the original text, supplied material, and were invaluable in discussing the subject matter. Sandy Van Osdol was particularly helpful in proofreading the revised text.

A note of appreciation goes also to Stanley Miller, J. William Schopf, Preston Cloud, Carl Woese, David Deamer, and Will Hargreaves for their assistance in correcting particular chapters in both the original and revised editions. These and others who contributed material for the book are credited in this volume.

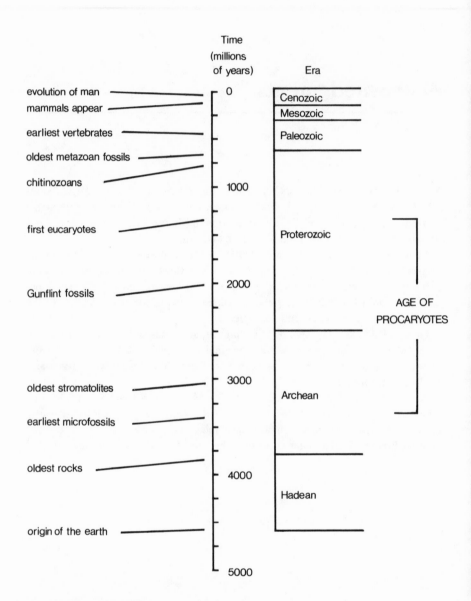

Geologic time and formations.

1
Building Blocks

The time was an autumn afternoon in 1951. Harold Urey, physical chemist and discoverer of deuterium, a heavy form of hydrogen, began his lecture at the University of Chicago on a subject that was one of his life's studies, the origin of the solar system. In his latest book,[1] then just recently completed, he had postulated that, because of the reduced conditions of the solar nebula that gave rise to the planets, the primordial atmosphere of the earth did not contain oxygen, as it does today, but consisted of methane, ammonia, and hydrogen.

The gathering of students and faculty sat listening with interest. Theories regarding the origin of the sun, the earth, and the planets had changed throughout the centuries. No longer was it believed that the planets were formed from molten globs of matter ejected from the sun during a close encounter with another star. In 1943, only eight years earlier, a German scientist had postulated that the planets were formed by the accretion of solid material in swirling eddies of an immense cloud of dust and gases. And only since 1929 had it been realized that the universe is mostly hydrogen and that the presence of free oxygen in the earth's atmosphere is a strange cosmic occurrence.

The professor went on into another favorite subject of his, the origin of life. Scientists generally believed that life must have begun on early earth billions of years ago in a manner that could be explained scientifically. But to demonstrate how it could have happened has been extremely difficult. As a result, studies on the origin of life were in a state of paralysis. The life sciences have shown that all forms of life consist of certain chemical substances that are the building blocks: amino acids, sugars, lipids, and two kinds of heterocyclic

1

bases called purines and pyrimidines. These components are linked together into large polymeric molecules: the amino acids form proteins; the sugars, polysaccharides; and the bases, nucleic acids. The assembly of these complex polymers into units bound by a lipid membrane creates living cells. The difficulty that thwarted scientists for over one hundred and fifty years, however, is that the building blocks of life appear to be produced in nature only by living organisms. There, then, lies the paradox. If only living plants and animals can synthesize the amino acids and other building blocks necessary for their own creation, how, Urey asked, could life ever have begun on earth?

Some students jotted down notes, most allowed their thoughts to sweep over the problem, searching for ready answers to the question. The seminar was becoming more engrossing. Theories on matter, energy, and the universe were fascinating and stimulating, but to speculate on the formation of life from inanimate substances excited the imagination and sensed of the momentous.

Urey went on: There must have existed circumstances on primordial earth that no longer exist that permitted life to begin. The absence of free oxygen and the reduced atmosphere would have produced a different chemical environment. In order for a living cell to develop, some organic compounds must have preceded the origin of life. But this was only a hypothesis. There was no experimental proof that organic substances necessary for the assembly of life could ever have formed on prebiological earth. Nonetheless, logic dictated that organic compounds must have been created on earth in some way before life could form.

As Urey continued to lecture on the paradox of life's beginning, one of the first-year graduate students, a young Californian by the name of Stanley Miller, listened intently. Miller, twenty-one at the time, had come to Chicago in September from Berkeley and was still searching for a suitable research problem to work on for his dissertation. The point that Professor Urey made seemed valid, but to prove how organic matter formed under primordial conditions probably would take a considerable amount of time. Miller brushed the thought aside and resigned himself to his original intention of finding a more theoretical problem.

The seminar ended with the customary question period. Did not the Russian biochemist Alexandre Oparin also discuss the possibility that organic compounds had been produced in a reduced atmosphere before there was oxygen? Urey answered in the affirmative, pointing out that no one had yet put the hypothesis to the test of experimental verification.

That winter the young graduate student from California discussed with the various chemistry professors their particular research interests in the hope of finding a rewarding project to work on for his thesis. Eventually, he decided to study under Professor Edward Teller, an authority on atomic physics, the manner in which elements are synthesized in extremely hot stars.

Six months later Teller announced that he was leaving the University of Chicago to set up a laboratory in Livermore, California. Faced with the problem of finding a new mentor and a new topic for his dissertation, Miller's thoughts turned to Urey's seminar and the questions it had posed. Perhaps the experimental work would not be as messy as he had originally imagined. The more he thought about it, the more the problem appealed to him and his enthusiasm mounted.

When he approached Urey with the proposition that he study how organic compounds could have formed on prebiotic earth, his enthusiasm was met with caution. The professor explained that the research could become a long and fruitless task. Perhaps Miller should consider studying the occurrence of thalium in meteorites. Only after realizing the student's determination did Urey consent to his attacking the problem of the synthesis of organic compounds. If, however, nothing came of the study within six months, he insisted that Miller switch to a more conventional research problem to assure success for his thesis.

During the following weeks Miller studied Urey's paper on the subject[2] and read the book by Oparin.[3] With Urey he then attempted to design a laboratory apparatus that would simulate the postulated atmospheric conditions of primordial earth. Something resembling a natural source of energy was needed to act on a mixture of gases to generate a chemical reaction. Since the atmosphere receives moisture evaporated from the oceans, which condenses again to return to the surface as rain, a supply of water would have to be included. A design was sketched and taken to the glassblower for construction.

A week later the drawing had been converted into a full-scale model of connecting glass tubes and flasks. The complete apparatus consisted of a 5-liter glass chamber atop a glass tube into which two tungsten electrodes protruded with their tips close enough to each other to permit a spark to leap across the gap. Joined below this was a condenser connected to a U-tube that ran over to a reflux flask for water; a return tube extended from the flask back to the chamber. It was a closed system in which water boiled in the flask would be carried as vapor past the electric spark, condensed, and returned to the flask. The apparatus was to be a model of the atmospheric conditions of primitive earth simulating the occurrence of thunderstorms in the primordial atmosphere.

Miller set up the apparatus and pondered its construction. He read the instructions for the Tesla coil that was to generate the electric spark and was astounded that it produced 60,000 volts. He hesitated, and doubts began to cloud his thoughts on the feasibility of the experiment. Hydrogen and methane both form explosive mixtures with air. Any leak could be disastrous. Even the thought of sparking 60,000 volts in water vapor seemed hazardous. And since it was an airtight system, heating the water to a reflux temperature would expand the gases and could create dangerous pressures. He took the apparatus back to the glassblower and had him interchange the positions of the condenser and the tube containing the spark gap.

With the construction now changed so that the spark came after the water vapor had condensed, it seemed safer, and Miller decided to put the experiment to the test. He added water to the flask, then pumped the air from the system and replaced it with a mixture of methane and hydrogen, making certain it was cleared

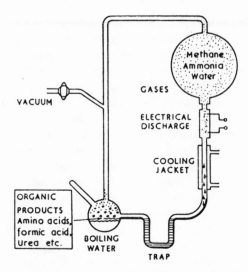

Figure 1.1. Apparatus used by Miller to simulate the prebiotic synthesis of amino acids and other organic compounds.

completely of any traces of oxygen. He checked for leaks and found the apparatus to be tight. Cautiously, he plugged in the Tesla coil and slowly ran the discharge up to 60,000 volts. Flashing streaks of blue darted across the gap between the tungsten electrodes with a crackling cadence.

Nothing else seemed to happen. Miller went to his desk and tried to study, but walked over to check the apparatus periodically that afternoon to see if he could detect any change. When it came time to go home, he decided to let it run overnight. The next morning when he entered the lab he saw a faint film of hydrocarbon floating on the surface of the water in the flask. There was nothing new in that. Earlier researchers had observed the same result when they exposed methane to an electric discharge. He allowed the experiment to continue for several days. The hydrocarbon film thickened, but an analysis of the water failed to show anything resembling organic substances of a biological nature.

Perhaps the 60,000 volts was less dangerous than had appeared. Miller returned the apparatus to the glassblower and asked him to restore the condenser and the tube with the electrodes to their original arrangement. After a week he was ready for another attempt.

The experiment was repeated. This time the water in the flask was warmed to a low heat with a heating coil. Again the experiment was allowed to run continuously. After two days Miller saw that the hydrocarbon film was no longer present and the solution had become a pale yellow. Something was happening. The water was analyzed, and the results were suggestive but not conclusive.

Again the experiment was repeated. This time Miller turned up the heat so that the water boiled vigorously. As he saw the water dripping from the condenser, he knew the apparatus had held the pressure until a steady state was reached and the water was circulating through the cycle. The discharge sparked and crackled across the gap as the boiling water drove its vapors with the gases past the electric discharge. Miller looked on with satisfaction as the experiment performed the way it was designed to in simulating the primordial atmosphere.

He could do nothing now but wait. Hour after hour the sparking continued as the water vapors and gases circulated around and around, imitating the cycle of water evaporating from the oceans into the atmosphere where it mixes with the atmospheric gases and is exposed to thunderstorms, only to return to the oceans as rain. For a long while there was no perceptible change. The experiment was left to run overnight.

The next morning when Miller entered the lab he immediately noticed that the water in the flask had turned pink. Excitement seized him and he rushed over to look more closely. Then as he saw that the heating mantle had an exposed coil that glowed red through the water, his exuberance was dampened. Slowly he lowered the mantle and looked at the solution in the flask. It was still pink. A definite chemical reaction had been taking place after all. His thoughts raced. Porphyrins? They give red color to blood. Were porphyrins being produced in this simulation of the conditions of prebiotic earth?

He let the experiment run. Day followed day. The color deepened. After one week the water in the flask had become decidedly red.

The time had finally arrived to test the results. Miller stopped the experiment and allowed the apparatus to cool. Extracting a sample from the flask, he analyzed it by paper chromatography, a standard procedure for separating and detecting small amounts of material. After running the chromatogram, he sprayed the strip of paper with a mist of ninhydrin solution and warmed it in an oven. Within minutes purple spots appeared, indicating the components. The compounds were amino acids!

Miller analyzed his one-gram sample by anion and cation exchange chromatography. He fractionated the components, made chemical derivatives of the ones in greatest yield, and compared the melting points to those of known amino acids. He sterilized the entire apparatus and repeated the experiment to be certain the results could not be due to bacterial contamination. There was no question of the results. They were amino acids, the very compounds that are used by plants and animals to construct their proteins.[4] No longer was there a dilemma of how organisms could have produced organic compounds before they themselves existed—the building blocks had already been there on primordial earth.

It was the experiment that broke the logjam. Its simplicity, the high yield of the products, and the specific biological compounds in limited number produced by the reaction were enough to show that the first step in the origin of life was not a chance event but one that had been inevitable. In effect, the experiment revealed that the basic components from which biological systems are constructed are energetically favored compounds. With the appropriate mixture of gases, any energy source that can cleave the chemical bonds will initiate a reaction resulting in the formation of life's building blocks.[5]

2
Early Earth

It was a stark, barren earth over which the sun rose quickly each morning, searing in a black sky in a blaze of intense ultraviolet radiation. The accretion of the unmelted mass of dust, aggregates, and stones which formed the planet had left it looking much like the dry, barren face of the moon. And as the sun followed its diurnal course, it rushed across the sky in a few hours to descend below the horizon just as quickly. For on this airless, waterless, hadean world, the day was only five hours long.[1]

Nightfall brought the rise of the moon, an awesome globe so close as to appear to touch the earth's surface as it loomed over the horizon, brightening the austere landscape with its huge glowing face. Each time as it ascended rapidly, its pull played in tidal oscillations on the viscous molten lava pouring onto the earth's surface from the many eruptions. The moon revolved just beyond Roche's limit of 2.86 radii (11,000 miles) and escaped the destruction into a ring system like the outer ring circling Saturn at 2.3 times the planet's radius. In quick succession the months passed, as the moon spun around the earth in 6.5 hours in an orbit approximately 46 degrees to the ecliptic.

As bleak as the rocky interior may have appeared, within its creation the inner earth contained the essence of a new existence. The planet was still essentially an unsorted conglomerate accreted at a temperature low enough to have retained the volatile constituents within its rocky structure. But also trapped in its interior were the radioisotopes of the elements potassium, uranium, and thorium. As time passed, the heat generated from the radioactive disintegration of potassium-40, uranium-235 and 238, and thorium-232, unable to escape, was absorbed by the rocks, and their temperature began to rise. After millions of years the accumulating heat drove the temperatures above the melting point of the silicates. As the rocks melted, they expanded and, becoming less dense, migrated upward while the denser material moved downward. A differentiation of the interior of the earth based on density, begun during the accretion stage, was now accelerated by a new dynamic mechanism.

Volatile components held in chemical bonds were liberated by the heat and created immense pressures beneath the weak, rocky surface. After a while the crust, no longer able to detain the cauldron within, opened in fissures, and the volatiles and molten material spewed out in volcanic eruptions, fumaroles, and hot springs. Oxygen, hydrogen, nitrogen, and carbon, long frozen in nonvolatile combinations with metals as oxides, hydrates, nitrides, and carbides, were freed from the nongaseous elements and blown and sweated to the surface, where they began forming an atmosphere. Any free oxygen quickly reacted with the reduced gases to form water, and the gaseous envelope that began to form over the globe consisted of hydrogen, nitrogen, water vapor, carbon monoxide, carbon dioxide, and the acids of sulfur and chlorine. Phosphorus, bound in rock as the mineral apatite, was freed by the intense heat and belched to the surface with volcanic ash to react quickly with water.

Beneath the hot surface, the cauldron began churning in great convection currents, moving the heat and lighter substances outward and the dense iron-nickel melt toward the center of the earth. The radioactive elements, bound selectively in the crystal lattice of minerals of less density, were carried toward the surface. There was no distinct crust at this time but only the outer surface of the yet unsorted conglomerate. The inner core, heated by the radioactive decay from above and the intense pressure of gravity, began to grow with the inward migration of the molten iron.

The surface was not stabilized as it is today by being of lighter material, and it continued to subside to displace magma that flowed to the top and poured out in great floods of lava. The molten rock that reached the surface solidified and took on the fine-grained texture of basalt, a volcanic rock coal black to dark gray in color, with the principal minerals being the ferromagnesium silicate pyroxene and calcium-bearing plagioclase.

The earth was an aggregation of rocky materials in which most of the rock-forming minerals were silicates of iron, aluminum, magnesium, calcium, sodium, and potassium. The oxides of iron were present, as were apatite, the calcium phosphate mineral. Only four elements—iron, oxygen, silicon, and magnesium—comprised 93 percent of the total weight. Most of the iron migrated toward the center of the earth to build the core, whereas silicates became the principal minerals of the outer layers.

Igneous rocks—those crystallized from the silicate melt called magma—are the closest approximation to the primordial earth material. Rocks may be a single mineral such as quartz (SiO_2), but generally they are a combination of minerals, the composition and physical properties characterizing the rocks. When magma

Table 2.1. Principal minerals in the earth's crust.

Mineral	Percent weight
SiO_2	60.18
Al_2O_3	15.61
CaO	5.17
Na_2O	3.91
FeO	3.88
MgO	3.56
K_2O	3.19
Fe_2O_3	3.14
TiO_2	1.06
P_2O_5	0.30

Source: F. W. Clarke and H. S. Washington, U.S. Geol. Survey, Profes. Paper 127 (1974).

solidifies at or near the surface, the slow growth of constituent minerals creates plutonic rocks, of which granite is a common type. Granite is one of the lightest rocks, having 66 percent or more silica; basalt has less than 50 percent; and andesitic rocks make up the intermediate range.

For hundreds of millions of years the earth was a hot, inhospitable place as volcanism poured out noxious fumes and vapors at an enormous rate. There were no oceans, a scant atmosphere, and a surface barren, pitted, and scarred by fissures and fiery eruptions from within. The earth would have seemed to have little destiny. But vast amounts of water bound in the rocks as hydrates were being liberated into the atmosphere and remained there as the surface was hot. After a very long time, with the air saturated and the surface of the earth cooling, a new phenomenon took place.

It rained.

It rained, and the rain evaporated, and it rained some more. It poured down on the bare rocky surface and ate the rock and collected in great flat basins. It was not the sweet rain of the earth's spring; it was the bitter, corrosive acid rain from the bowels of the earth, heavy-laden with hydrogen sulfide, carbon, and hydrogen chloride. The principal volatiles from the volcanoes were water, carbon dioxide, and hydrogen chloride in a ratio of 20:3:1; the rain was approximately 1 molar hydrochloric acid.

But as the rains were acid, bringing with them the chloride, bromide, sulfide, and carbon dioxide, the rocks were basic with sodium, potassium, and calcium. The acid rain dissolved the rock until it was neutralized, and where the water evaporated the salts formed broad, flat salt plains.

As the volcanoes continued the outgassing of the earth's interior, they were restoring the atmosphere and creating oceans. It was a reducing atmosphere devoid of oxygen, and the oceans were merely shallow catchbasins for gathering the rains. Nearly two billion years would pass before oxygen would be present in

Table 2.2. Volatile materials now on or near the earth's surface unaccounted for by rock weathering.

Volatile	Wt. (10^{20} grams)
Water	16,600
C as CO_2	910
Sulfur	22
Nitrogen	42
Chlorine	300
Hydrogen	10
B, Br, Ar, F etc.	4

Source: Modified from W. W. Rubey, Bull. Geol. Soc. Am. *62* 1111–1147 (1951).

significant quantities, and the oceans came into being only by growing throughout the ages from the water expelled from the earth's rocky interior.

Two billion years in the future the oceans would achieve their modern characteristics (see chart, p. xx). But the oceans of Hadean Earth were solutions resulting from an acid leach of basaltic rock. The atmosphere was devoid of oxygen, so that anaerobic deposition environments with carbon dioxide pressures of about $10^{-2.5}$ atmospheres, or ten times today's level, were prevalent. Under these conditions, the pH* of early ocean water was lower than today's. The calcium concentration was higher, and the ocean was probably saturated with respect to amorphous silica. In addition to the other ions from the basaltic rocks, reduced iron and sulfur would have been in their proportions found in the rocks. Only when the pH approached neutrality would aluminum ions begin to precipitate as the hydroxide and combine with silica to form cation-deficient alumino-silicates. As long as the hydrogen chloride exceeded carbon dioxide, the oceans would have had a high content of calcium chloride, and the carbonate would not have been precipitated, as it was at later times.

As the atmosphere was being formed, important changes began to take place. In the upper level, radiation from the sun dissociated water molecules to hydrogen and oxygen. The hydrogen escaped to outer space and the oxygen reacted quickly with the reduced atmospheric gases, reverting to water. The photodissociation continued to consume some of the atmospheric water, but as the principal volatile brought to the surface by volcanic activity, it accumulated much faster than it was consumed. As carbon dioxide poured into the atmosphere from the volcanoes, the amount built up was restricted by absorption of the carbon dioxide into the oceans. In this way the level of atmospheric carbon dioxide was kept at a relatively low level.

---

* pH is a logarithmic scale of hydrogen ion concentration. Neutral solutions have pH 7; 0–7 is acidic; 7–14 is basic.

It has generally been believed that the earth heated to a hot stage at this time, forcing out the water and other volatiles into a dense atmosphere, from which they would condense out at a later time after the earth cooled sufficiently. William Rubey,[2] who studied the matter, has found convincing evidence that this could never have happened. Instead, it appears that at no time in the earth's history has there ever been more than a fraction of the excess volatiles of the earth's interior in the atmosphere.

The amount of carbon dioxide buried as carbonates and organic carbon in sedimentary rocks is 600 times greater than all the carbon in the atmosphere, hydrosphere, and biosphere. If even as much as 1 percent of the carbon dioxide now locked in these rocks had been in the atmosphere, the oceans of today's volume would have gone from their pH of 8.2 down to 5.9.

The equilibrium of atmospheric carbon dioxide and that absorbed by the oceans proved to have profound significance. Carbon dioxide in the atmosphere creates what is called the greenhouse effect. Like the glass of a greenhouse, carbon dioxide in the atmosphere is transparent to visible light but absorbs heat-generating infrared rays. When sunlight is absorbed by the surface of the earth, the warmed solid substances reemit much of the energy as invisible infrared radiation. If the level of carbon dioxide in the earth's atmosphere were too high, this energy would be reabsorbed by the air instead of being radiated into space, and the earth would heat up. By absorbing carbon dioxide and keeping the atmospheric concentration level low (about 0.024 percent today), the oceans have controlled the greenhouse effect on the earth.

On Hadean Earth there were no continents, and the crust was the outer surface of the mantle. The radioactive elements were not yet concentrated in the crust or upper mantle as they are today, but were still distributed throughout the undifferentiated mantle. Nonetheless, the accumulating heat from the nucleoclides over hundreds of millions of years was fractionating the earth into concentric layers. And as the radioactive elements migrated upward toward the outer crust, they brought with them their heat-producing capacity nearer the surface. The magma would have concentrated in a transitional layer below the surface as it is today between the crust and the upper mantle.

Toward the end of Hadean times the buildup of radiogenic heat, the partial melting of the mantle, and the upward convection of heat in the magma must have reached climactic proportions, causing the original crust to undergo modifications in areas around the globe. The intrusion of igneous rocks and the extrusion of lavas deformed the surface and accelerated erosion, and the earliest sedimentary rocks formed.

It was probably still too early for the formation of true crystalline rocks. Nor did the sea-floor spreading process—in which crustal plates separate by the welling up of magma—occur during this time. There was not yet the irreversible differentiation of the initially homogeneous mantle. But the passing of the Hadean time was marked by the release of the heat buildup in great floods of lava spewing out from volcanoes onto the earth's surface and beneath the waters of the growing oceans. It was the end of the first long age of the earth's history, an era that had seen the birth of the earth's atmosphere and the oceans; it had lasted 800 million years.

Still 3.8 billion years in the future the rocks crystallized from the magma in that primordial scene were to play a significant role in science. In 1966, Vic McGregor, a young geologist from New Zealand working with the Geological Survey of Greenland, began mapping in detail the mountain area around a fjord named Amerlik on the western coast near Godthaab, the capital. It was not an easy task. The great assortment of rocks and their complex arrangements made interpreting the geologic history difficult. But after several years McGregor began to piece together a distinctive sequence of events that had occurred down through the ages. The rocks recognized to be the oldest from that region were the Amitsôq gneisses, igneous rocks that had undergone metamorphosis and deformation by mighty forces acting on the earth's crust. By McGregor's interpretation, there should be rocks that were formed even earlier than the Amitsôq gneisses.

Stephan Moorbath of Oxford joined McGregor in the summer of 1971, and the two geologists began to collect rock samples to be sent back to England for age measurements by radioactive isotopes. At Isua, a mountainous area 60 miles northwest of Godthaab at the very edge of the great inland ice sheet, a mining company was exploring a large iron ore deposit. The ore was a part of a great arc of highly metamorphosed volcanic and sedimentary rock 7 to 15 miles in diameter and 1.8 miles thick. When McGregor and Moorbath reached the site, they saw that the arc of rock was supercrustal, that is, laid down on a surface and flanked by granite gneisses with edges of contact sheared and deformed.

Samples were collected and sent back to Oxford for rubidium-strontium and uranium-lead measurements to determine their ages. When the results were finally received, the age of the rocks was found to be 3.76 billion years. McGregor and Moorbath learned that on that bleak Arctic upland near the great ice sheet where the rock is totally exposed, they had been walking on a section of what may have been the original continental crust of the earth.[3]

Throughout the Hadean era the internal heat distilled to the surface the volatile materials bound to the rocks, and a primitive atmosphere and hydrosphere were born. By the time of the Isua Iron Formation 3.76 billion years ago, there was enough water on the surface for rain, erosion, and sediments. Following the Hadean, the earth entered the Archean era.

The early Archean would have been a consolidation interval following the great eruptions that ushered out the Hadean. Over the hundreds of millions of years erosion wore down the old volcanic mountains and deposited thick layers of sediments on their margins. While the surface was being weathered, within the earth's interior the heat of radioactive decay continued to accumulate on a worldwide scale for the next episode.

Then around 3 billion years ago, when shallow primitive oceans skimmed the earth's surface, the earth underwent another period of crustal formation. The cauldron no longer detained, the crust split into huge rifts, and the molten magma caused the adjoining sections of crust to decouple from the upper mantle and slide over the layer of molten rock at the interface. The movement wasn't much— only a centimeter or so a year—but over the millions of years it was enough to force the surface of the earth to lose this increase in its crust in some manner elsewhere. In order to relieve the pressure the crust split at other places and the section under pressure began to slide over the neighboring section, driving down the edge of cleavage into the mantle beneath it.

As the subducted surface slid deeper into the mantle, it melted at the subterranean temperatures. The lighter silica-rich rock moved upward, while the ferromagnesian rocks sank. Where one edge of the surface slid beneath the other, a deep trench resulted, and parallel to the trench the subducted crustal material extruded as magma onto the surface, creating an arc of volcanic islands. These island arcs must have appeared in a number of locations on the face of the earth.

The magma that erupted along rifts in the earth's crust, pouring out vast quantities of lava, became the basement rocks of the continental shields. These foundation rocks, although 3 billion years old and folded and metamorphosed, can be seen as long successions of pillow lava as much as several miles thick, formed by the rapid chilling of being erupted under water. There is evidence that these belts, known as greenstone belts because of their greenish tinge from chlorite, hornblendes, and epidote, may have been crustal regions that subsided 6 to 9 miles as volcanic rocks and sediments accumulated layer upon layer over a long passing of time.

The rocks in greenstone belts have no exact equivalent among contemporary active volcanic regions, although they show some resemblances to present volcanic island arcs, such as the Kuriles north of Japan and the southwestern Aleutian islands. The rocks around the greenstone belts are often predominantly granite, and many Archean provinces consist of strongly folded, steeply dipping volcanic successions compressed between granitic bodies. The distribution of land and water during the Archean is not well known, but the sediments and volcanic rocks were accumulating under water.

The island arcs then were probably what they appear to be today—a significant evolutionary intermediate stage between oceanic and continental crusts. Oceanic crust is basaltic and represents the surface of the mantle, whereas continental crust, thicker and of lower density, is principally granite. Island arcs are of crustal material thicker than oceanic crust, but not as thick as continental, and composed of rocks common to both.

Toward the end of the Archean another stage of crustal development occurred. As oceanic crusts slid beneath the overriding plates, the molten rock from the subducted crusts rose by its buoyancy and moved into placements on the margins of the island arcs. Sometimes the magma extruded onto the crust, pouring out thick series of volcanic rock. More often, it remained intruded until the final

Figure 2.1. The growth of North America. The numbers indicate the ages of rocks of the continental platform in billions of years.

stage of mountain building, or orogeny, as it is called by geologists. Orogeny went on over hundreds of millions of years. Then, at a time when the confining pressure over the intrusive body weakened, its buoyancy drove it upward like a cork through the crust in the last, spectacular stage of mountain building.

In this way the continents have grown step by step from their nucleus of granite by addition to their margins of the lighter material fractionated from the mantle. The entire procedure of mountain building apparently has occurred in six or seven episodes throughout the earth's history, each episode lasting approximately 800 million years.

The thrust of the mountains on the margins of the greenstone belts severely metamorphosed and deformed the rocks. The subsidence of the belts created the flat, low-lying terrain known as the continental shields. These are the cratons, or nuclei, from which the continents grew by subsequent mountain building episodes. The Canadian Shield, the largest of these features, lies like a gigantic saucer with the center holding the Hudson Bay. The mountains are gone now, long ago worn down to their root stocks, but the area remains a vast hoard of mineral deposits.

The greenstone belts, dating from 3.4 to 2.5 billion years ago, are the oldest belts of metamorphosed and deformed rocks. From about 2.7 billion years onward, individual sedimentary basins formed from the erosion of the ancient mountains. The cratons upon which the continents were growing may have been 5 to 10 percent of the present continental area 3.8 billion years ago, but between 2.9 and 2.5 billion years ago they grew in area until they were 50 to 60 percent of the present area. The earth's crust was becoming stable enough by 2.7 billion years ago to allow sediments to accumulate in large basins without being altered by subsequent stresses.

During the Archean the atmosphere was still without oxygen, but it contained nitrogen, and the carbon dioxide content was probably 4 to 10 times today's value, keeping the lower atmosphere extremely hot. According to geologists L. Paul Knauth and Samuel Epstein,[4] in studies at the California Institute of Technology using isotopic analysis of 66 samples of chert from central and western United States, 3 billion years ago the average temperature may have been as high as 70°C (160°F). The findings are based on measurements of the relative abundances of isotopes of oxygen and hydrogen in water of hydration of chert from different past geologic ages. The data indicate that climatic temperatures with some fluctuations have been generally declining from that time.

The Archean era lasted until 2.5 billion years ago. When it ended the crust was stable enough to hold heavy platforms of sediment and sufficiently rigid to withstand intrusions of magma. From then on the geologic column is characteristically cratonal sediments deposited on submerged continental margins.

The Archean era was followed by the Proterozoic—an era that extended for nearly 2 billion years. It came to an end 570 million years ago with the beginning of the Cambrian period—and fossils.

3
Life before the Precambrian

The work of constructing a standard stratigraphic and geologic chronology was vigorously pursued by geologists from the early part of the last century. As their science expanded, they learned that unique fauna were associated with particular intervals in geologic time. By integrating the sedimentary rock throughout the world into an orderly sequence based on faunal succession, they were able to construct from fossils a geological column for the history of the earth.

Adam Sedgwick, an English geologist, was the first to use the name Paleozoic, in a lecture in 1838, to designate the era of geologic history from the earliest fossils to and including the land plants, amphibians, and the earliest reptiles. Soon Mesozoic was used for the era of dinosaurs and marine and flying reptiles, and Cenozoic for the most recent era. The Paleozoic, Mesozoic, and Cenozoic eras were subdivided into periods. And the Cambrian period, set at the first appearance of fossils 570 million years ago, was followed in succession by geologic periods characterized by their fossils.

The fossil record, though fragmentary in some instances, gives clear testimony to the principle of evolution. In many lines of descent, the sequence can be worked out in considerable detail. The most ancient fossils from the Cambrian include only invertebrates. Later, somewhat fishlike vertebrates appear and gradually blend in succession to true fish. Fossils of amphibians and reptiles follow, and finally birds and mammals. The geologic column is clearly correlated with the simplest form of animals appearing in the oldest strata and evolving to greater complexity late in geologic history.

As we study the fossil record, we find that the Cambrian fauna contain representatives of every important invertebrate phylum, but the only groups that are abundant and widespread are the trilobites and the brachiopods. Almost 75 percent of all fossils found in the Cambrian are trilobites. These arthropods, which were distantly related to the modern horseshoe crab, ranged in size from one-fourth of an inch to almost two feet and fed on microscopic organisms in the sea and on the bottom detritus. For the next 70 million years of the Cambrian period, from 570 to 500 million years ago, there is no record of any vertebrate fossils or any plants and animals on land or in fresh water.

The abrupt appearance of the Cambrian fauna constituted a major biological problem. The various organisms found in the Cambrian are life forms with organs and characteristics as complex and developed as those of some found today. Furthermore, to compound the mystery, all phyla (divisions based on fundamental anatomical features) known today are present in the Cambrian fossils. All, that is, except one: the Chordata, the phylum of the vertebrates. The vertebrates were not to appear until about 450 million years ago. The phyla that were present appeared quickly with no apparent origin. Moreover, the earliest Cambrian beds with skeletal remains contained several distantly related trilobite, brachiopod, mollusk, and echinoderm groups and representatives of the other phyla that totaled about twenty distinct kinds of invertebrates, but there was no indication of convergence back toward a common ancestor. Where the Cambrian plants and animals came from remained a mystery.

Then in 1947, Reginald Sprigg, an Australian geologist, was exploring the Ediacara Hills, an abandoned mining area 380 miles north of Adelaide, when he discovered abundant fossil remains of jellyfish in upper strata of quartzites. At first he believed them to be of the lower Cambrian, but further investigation established the rocks to be late Proterozoic. About 1,500 fossil specimens were collected from the beds. Two-thirds of these fossils were outlines of the characteristic swimming bell of the coelenterate medusa, around one-quarter were annelid worms, and the remainder, extinct invertebrate organisms.[1]

The composition of the fauna was that of a marine environment, and studies of the encasing sediment indicated their deposition in shallow water. The wormlike creatures probably lived in shallow water, where they tunneled in the mud or fed on the surface, whereas the *Medusa* may have drifted in from the open seas. There were no signs of predation, such as the tearing of large bodies.

Some of the members of the Ediacara fauna have been found elsewhere. *Charnia* were discovered by Trevor Ford in England in Proterozoic rocks with an age of 680 million years. And a fossil closely resembling *Charnia* was found in the Olenck Highlands of northern Siberia in rocks dated at 675 million years.

The fossils of many of the animals of the Cambrian period were invertebrates with mineralized skeletons. These organisms were absent in the Ediacara fossils, which yielded impressions of soft-bodied animals or tracks and trails of invertebrates without hard parts. There appears to have been a period of possibly 100 million years when soft-bodied animals flourished in the sea, as many do today, before the evolutionary development of mineralized shells and skeletons.

The significance of the Ediacara fauna is that the animals belong to a phylum that is on a simpler level of development than those phyla found in the Cambrian. These animals are coelenterates, represented today by jellyfish, sea anemones, and corals. The coelenterates are multicellular animals at the tissue level of construction, which means in general that they lack organs.

The only multicellular animals simpler in form than the coelenterates are sponges. These primitive animals lack a well-defined organization of tissue and indicate similar ties with certain types of protozoan colonies, both lacking integrated parts, mouth, and digestive systems, both having a type of skeletal formation in which single elements are produced by a single cell or by a group of cells. Sponges are not well preserved as fossils, but specimens from the Cambrian have been found.

The measurement of fossil-bearing strata carried the age of life back 570 million years to the base of the Cambrian, and the Ediacara fossils extended this to 680 million years. In no place on earth, however, are the strata all laid in one continuous succession. If the maximum strata for all the ages from the time of the earliest fossils were stacked vertically, the column would be nearly 400,000 feet (76 miles) thick.

Yet, it was apparent that the fossil record seriously underestimated the age of the earth. Beneath the Cambrian lie the pre-Cambrian formations—strata of volcanic and sedimentary rocks of immense thickness and covering a time span of over 3 billion years: five times the length of time since the earliest fossils. What happened in that incredibly long span of time representing nearly 85 percent of the earth's history before the Cambrian phyla inhabited the earth?

The enormous thicknesses of Cambrian strata attest to the millions upon millions of years that life existed only on the level of sponges, jellyfish, and trilobites. In the Inyo Mountains of California the strata containing trilobites and archaeocyathids, an extinct type of calcareous sponge, descend for 14,000 feet, or nearly 3 miles. Yet, this is not down to the point when life first appeared. There is tangible evidence that living organisms existed on earth for a very long time before the Cambrian period.

The discovery of the Ediacara fossils and others found below the Cambrian strata that began 570 million years ago caused a restriction in the definition of Cambrian rocks to those strata that contain fossils recognized to be characteristically Cambrian. As more geological surveys were conducted, it became apparent that the Cambrian fossils emerged from a trail that led back deep into earlier periods. In Morocco it was found that 3,000 feet of archaeocyathid-bearing strata conformably underlie the lowest Cambrian, and in turn, rest upon 10,000-foot-thick limestone containing the ancient remains of "water biscuits," the calcareous masses of fossiliferous mats and tufts with concentric laminations that grew in shallow water.

These cabbage-shaped or branched laminated structures called stromatolites are the most widespread and abundant fossils before the Cambrian. They are generally composed of limestone or dolomite, but in some cases they are sili-ceous. Stromatolites were probably formed in an intertidal environment in the same manner that they are still being formed today, by cyanobacteria, or blue-green algae. Many of these hemispherical fossils are as small as buttons, but others can be thousands of feet in area. The giant stromatolite domes in the Belt Series near Helena, Montana, are up to 15 feet thick and extend for thousands of feet.[2]

Although stromatolites were formed in greatest abundance during the long Proterozoic era, their occurrence extends back into the Archean era, where they have been found in the African Pongola Formation dated at around 3 billion years. Like most fossil stromatolites, those of the Pongola show no cellular detail. Nevertheless, they resemble structures of the type being formed today in the Bahamas and at Shark Bay, Western Australia,[3] and well-preserved stro-matolites of various ages have yielded microscopic cellular detail. No inorganic processes are known which form such structures.

Stromatolites are not the only evidence that the cyanobacteria must have an extremely old ancestry. Additional testimony is the oxygen. The earth's atmo-sphere with 21 percent free oxygen is an oddity in a universe that is 75 percent hydrogen. The oxygen-liberating photosynthesis has created an oxidizing at-mosphere that has elevated all life to a thermodynamically unstable situation.

Oxidation of organic compounds is a spontaneous reaction, which means that without a continuous input of energy, all biological matter would ultimately revert to the oxidized state of carbon dioxide and water. This unstable condition is maintained by the absorption of energy from sunlight for the reduction of carbon dioxide and the liberation of oxygen, 90 percent of which is generated by the planktonic algae of the oceans. It has been estimated that if photosynthesis were to cease today, all the oxygen in the atmosphere would disappear within 2,000 years through absorption by rocks unsaturated with respect to oxygen.[4]

Despite the earth's losing its protoplanetary atmospheric hydrogen, the at-mosphere formed by outgassing was also reduced and devoid of free oxygen. For photosynthetic organisms to oxidize the constituents of the atmosphere and hydrosphere before the buildup of free atmospheric oxygen was possible must have taken an extremely long time. But there is geological evidence that free oxygen began to accumulate in the atmosphere as long ago as 2 billion years. Between 2.3 and 2.0 billion years ago there was the last apparent occurrence of abundant and easily oxidized detrital pyrite and uranite.[5] These minerals, which were eroded from rocks and transported considerable distance before being laid down in sediments, would have been oxidized if the atmosphere at the time had contained significant levels of oxygen. The fact that they do not occur in later sediments as detrital deposits in abundance suggests that atmospheric oxygen was beginning to build up at this time.

And between 2.2 and 1.8 billion years ago, virtually the last and volumetrically the greatest phase of deposition of banded iron formation took place.[6] The

the greatest phase of deposition of banded iron formation took place.[6] The reducing conditions of the Archean and Early Proterozoic favored the formation of minerals containing ferrous iron from the alteration of basaltic rocks. The characteristic minerals that occur in iron formations are siderite (iron carbonate), greenalite (iron silicate), and pyrite (iron sulfide) in association with chert, originally an amorphous silica. But much if not most of iron formation was biogenic. Ferrous salts are relatively soluble, whereas oxidized iron is not. The liberation of oxygen by photosynthetic organisms oxidized the ferrous iron, creating deposits of precipitated iron. These became banded iron formations, unique rocks consisting of alternating layers of iron-rich and iron-poor silica representing rhythmically banded precipitated sediment.

It is uncertain whether the cherty iron zones of the 3.76-billion-year-old Isua Formation of west Greenland[7] or the 3.4-billion-year-old Onverwacht Group of Africa resulted from biological activity. The last major episode of banded iron formation was 1.8 to 2.0 billion years ago. Thereafter, with much or most of the iron oxidized, the oxygen was added to the atmosphere. Iron deposits that have formed in more recent geological time are red beds where the individual grains are coated with ferric oxide.

The paleochemistry of oxygen can thus be traced back to the pre-Cambrian red beds and limestones. From that time too "coals" containing beds of almost pure carbon have been found in Michigan and Finland. It is difficult to explain these except that they apparently were formed by well-organized photosynthetic life. As impressive as is some of the evidence of early life forms, it is circumstantial in that these are products of life before the Cambrian—not fossils of the actual organisms.

In the early 1950s, however, Stanley Tyler of the department of geology at the University of Wisconsin was prospecting for iron along the Michigan shores of Lake Superior when he came upon ancient coal deposits. The coal contained what he thought to be microscopic plants. Tyler showed the coal to William Shrock, the chairman of the geology department at the Massachusetts Institute of Technology. Shrock thought the plants looked like the fungus that grows on a jar of jelly that has been left open too long and suggested to Tyler that he show them to Harvard botanist Elso Barghoorn.

The result of this consultation was that Tyler and Barghoorn teamed up and made a field trip to the site for a closer study. A search for whatever had wrinkled the seam of coal led the two to the Canadian side, where they found black shales and chert known as the Gunflint Formation, a layer of Precambrian rock on the northern shore of Lake Superior near Shreiber, Ontario, just east of Thunder Bay. The Gunflint chert is overlaid by shale and is generally regarded as Middle Hurian of the Canadian Shield.

Samples of the chert were collected and cut with a diamond saw into slices so thin that light could pass through them. When Barghoorn viewed the thin

Tyler and Barghoorn cut more than 800 thin sections of the flintlike black chert for study. Hydrofluoric acid was used to dissolve the silica of the bedded chert to free fragments of the primitive plants and the organic residue of spores and filaments. Five morphologically distinct forms of biota were recognized: two were algal, two fungal, and one appeared to be calcareous flagellate. These tiny, very simple plants seemed to be representative of cyanobacteria and simple forms of fungi. The Gunflint Formation was dated at 1.9 to 2.0 billion years ago, making these fossils at that time the oldest structurally preserved organisms.[8]

In a subsequent paper in 1965, Barghoorn and Tyler[9] reported finding an array of microscopic fossils in other samples of the Gunflint Formation. In this paper they show 12 assemblages of unicellular spheroidal and filamentous microfossils. From their morphology the spheroids appeared to be related to the coccoid blue-green algae, and the filamentous fossils to the modern iron bacteria *Sphaerotilus* and *Siderococcus*. These were all excellent specimens, preserved by the mineralization of the cellular structure in a siliceous matrix. Here, entombed in 2-billion-year-old chert, were the remains of the microorganisms that had precipitated the ironstone formations and generated the free oxygen that made future life possible.

Soon thereafter, J. William Schopf,[10] a former graduate student of Barghoorn and presently at the University of California, Los Angeles, found varied assemblages of excellently well-preserved spheroidal and filamentous plant microfossils in the Bitter Spring Formation of Northern Territory, Australia. These fossils of microbiota occurred in carbonaceous cherts of the Late Precambrian strata in the Ross River area of Central Australia and were thought to be approximately 1 billion years in age. The fossils resulted from algae that apparently grew as laminar sheets or mats in a marine environment, forming widespread algal stromatolites. Of 19 species that Schopf found, 14 were of modern algal families. The filamentous and coccoid blue-green algae were predominant and must have been highly diversified at this time 1 billion years ago.

Apparently these cyanobacteria were flourishing as early as 2 billion years ago and may well have been responsible for the Bulawayan stromatolites of Zimbabwe, which formed 2.6 billion years ago, and the Pongola stromatolites of 3 billion years. However simple these plants were, they would have had predecessors that existed even earlier—microorganisms that were more ancient than the oldest blue-green algae. A goal of paleontology became a search to push back the fossil record as close as possible to the moment when life began on earth.

In 1965, Barghoorn was collecting cherts from several localities in the Barberton Mountain Land of the area of eastern Transvaal in South Africa near the Swaziland border where the waters of the Umbilizi River flow through the rolling hills on their way to Mozambique and the Indian Ocean. The Barberton Mountain Land is a few hundred square miles of hills formed of Archean greenstone belts whose ages date back as far as 3.4 billion years. The Swaziland Series of the formation consists of the Fig Tree Group underlain by the Onverwacht Series. Carbon is widespread in the formation, and in some places shales that were converted to graphic schists by metamorphism are found. Some black cherts contain as much as 0.20 percent organic matter.

Barghoorn and Schopf examined rocks of the Fig Tree Group by light microscope. The rock matrix contained many laminations of dark-colored, nearly opaque particles of organic substances. The laminations being aligned parallel to strata of chert suggested deposition in an aqueous environment. Nothing resembling fossils of microorganisms could be seen by the microscope. But then Schopf polished the surface of the rock sections and viewed them through an electron microscope. Under the greater magnification it was possible to see what had not been seen before. There were rodlike structures 0.5 to 0.7 millimicron long and 0.2 millimicron in diameter. They resembled rod-shaped bacteria! Later, spheroidal microfossils 17 to 20 millimicra in diameter resembling modern blue-green algae of the coccoid group were also found. In all, 29 well-defined specimens were detected in the fossils of the microbiota. These were fossils of what appeared to be bacteria that had been living on earth 3.0 to 3.3 billion years ago.[11]

At the same time, Hans Pflüg[12] of the Justus Liebig University of Giessen, West Germany, was examining cherts and shales of the Fig Tree sediments for structured organic remains. In samples collected in the vicinity of the Sheba Gold Mine near Barberton he found assemblages of remains of organisms. Chemical and optical studies of the organic material showed these structures to have cell walls. Pflüg suggested a similarity to ocalean cyanobacteria. The radiometric dating placed the age at 3.2 billion years.

These microfossils from the Fig Tree Group have to be regarded almost certainly of biological origin and are probably remnants of single-celled, algalike microorganisms. The organic composition, the constant morphology, the limited size range, and the similar appearance to the excellently preserved microfossils of the Gunflint cherts and the blue-green algae of the Bitter Spring Formation are strong indications of a unicellular, noncolonial, algalike form of life on earth over 3 billion years ago.

Approximately 35,000 feet below the Fig Tree Group lie the Onverwacht Series, covering 400 square miles of the southern part of the Mountain Land and 50,000 feet thick. In 1968, A. E. J. Engel and others[13] reported finding cup-shaped and spherical microstructures in this group of Archean rocks. The size of 590 structures ranges from 6 to 193 millimicra with no dominant size. This variance of 30-fold appeared, however, to be too much of a spread to be characteristic of organized biological systems.

Following this report, Jim Brooks and Marjorie Muir[14] investigated specimens of the Onverwacht strata by first treating slices of uncrushed rock with 20 percent hydrofluoric acid to digest the inorganic matrix. They recovered a concentrated dark residue of organic material representing 0.2 to 0.48 percent by weight of

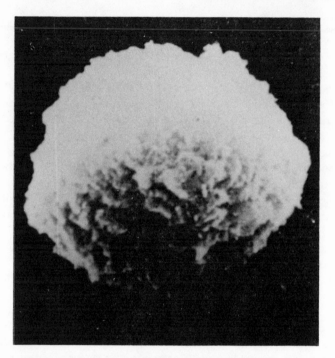

Figure 3.1. Microfossil from the Onverwacht Formation.

the samples. When viewed with an electron microscope the organic material was seen to be fossil remains of what appeared to be cell walls of microorganisms. There were two basic types: spheroids 7 to 10 millimicra in diameter and filamentous forms 15 to 20 millimicra in size. These fossils of microorganisms showed morphological similarities with those of the overlying and younger Fig Tree Group.

Chemical extraction and analysis of the organic matter from the Fig Tree Group have shown 0.003 to 0.015 parts per million of aliphatic hydrocarbons (C_{15}–C_{25}) and larger amounts of pristane ($C_{19}H_{40}$).[15] The Onverwacht chert had free aliphatic hydrocarbons, fatty acids, n-parafins C_{12}–C_{24}, pristane, and phytane.[16] Pristane and phytane, being isoprenoid hydrocarbons, are indicative of biogenic origin. But Nagy,[17] examining the porosity and permeability of the Onverwacht chert, showed that the hydrocarbons could have percolated into the rocks from above.

Nevertheless, most of the organic material in the rocks was kerogen, an intractable and insoluble residue that would have been formed in place. The kerogen was analyzed in a different manner by Dorothy Oehler[18] while she was working on her thesis with Schopf at UCLA. Photosynthetic organisms show a preference for $C^{12}O_2$ over $C^{13}O_2$ when they absorb carbon dioxide. By measuring the C^{13}/C^{12} ratio in kerogen, it was possible to establish whether or not it came from photosynthesis. Oehler's results suggested that it did and that autotrophs capable of fixing CO_2 had been on earth more than 3 billion years ago.[19]

Using the carbon isotope method, Schopf and his students[20] applied it to study the oldest stromatolites then known—those of the Bulawayan Formation. They confirmed that these structures were created by photosynthetic organisms, and Schopf suggested that their age of 2.6 billion years might be minimal for the time of origin of blue-green algae or photosynthetic bacteria as filamentous, integrated biological communities. The blue-green "algae," or at least the pho-tobacterial ancestors of all algae today may go back as far as the earliest known fossils of living organisms on earth, those of the Onverwacht Series and the morphologically similar microfossils discovered in the Warrawoona Group at North Pole, Western Australia, and found to be 3.5 billion years old.[21]

Similar microfossils from the Archean and Proterozoic resembling bacteria and blue-green algae have also been found in Ontario, eastern California, south-ern Africa, central and southern Australia, and the USSR.[22] But there is a degree of uncertainty about organized structures over 2 billion years old. Organic ma-terial will commonly aggregate into spheroids, so this alone is not sufficient in establishing biogenic origin. Of the fossil-like microstructures more than 2 billion years old that have so far been described, many were probable not fossils of organisms.[23] Nevertheless, stromatolites of probably algal origin were flourishing as early as 3.0 to 3.1 billion years ago, and the algae responsible for them could be related to forms seen in Fig Tree and Onverwacht sedimentary rocks and those of the Warrawoona Group.

These possible microorganisms of the Archean are assumed to have been procaryotes, the simplest known form of living cell. At some time in the pre-Cambrian era the eucaryotic cell, the more sophisticated and complex type of biological cell that gave rise to all later forms of life, had to have appeared. If Preston Cloud, a geologist at the University of California, Santa Barbara, is correct, eucaryotes emerged between 2.0 and 1.3 billion years ago.

In 1966, Cloud and his coworkers collected samples from black chert found 18.5 meters below the upper contact of the Beck Spring Dolomite in eastern California. Study of thin sections revealed preserved unicells and spiny sporelike bodies. These fossils were a line of microscopic spherical and filamentous shapes not unlike older fossilized microorganisms—except for two important differ-ences: they were much larger than older forms; and some of the filamentous forms were branched.[24] Recent analysis of microfossils for size and distribution shows that eucaryotes are generally about ten times larger than the procaryotes.[25]

Fossiliferous outcrops occur 2,900 meters below the lowest metazoan trace fossils and are younger than 1.7 billion years. These have been correlated to another group dated at 1.2 to 1.4 billion years of age. The most numerous fossils from the various localities are filamentous cyanobacteria placed in the new genus, *Beckspringia*. By studying the microflora in formations of determined radiogenic ages, Licari and Cloud have attempted to bracket the origin of eucaryotes between about 1.3 and 1.6 billion years ago.[27]

A dispute exists as to when the eucaryotes appeared. Helen Tappan[28] of the University of California, Los Angeles, believes they existed at the time of the Gunflint Formation 2 billion years ago. Knoll and Barghoorn,[29] on the other hand, deny the existence of fossilized cell organelles and the appearance of eucaryotic cells before the *Metazoa*. If Preston Cloud, Gerald Licari, and others are correct about the Beck Spring Dolomite fossils, however, these fossils will then mark the greatest biological breakthrough in the history of life on earth—the eucaryotic cell.

The fossil record is still sketchy for the transitional period from the unicellular microorganisms of 1 billion years ago to the Ediacara fauna 680 million years ago, when the *Metazoa,* or multicellular animals, emerged with cells organized in layers of tissue. A discovery in the 1970s, however, may have helped to bridge the gap. Bonnie Bloeser and Schopf of UCLA, Robert Horodyski of Tulane University, and William Breed of the Museum of Northern Arizona[30] reported finding microfossils in the pre-Cambrian rocks of the Grand Canyon which appear to belong to a distinctive group of one-celled planktonic organisms called chitinozoans having an age of 750 ± 100 million years. The chitinozoans are thought to be unicellular heterotrophs, which means that they, like all animals, cannot produce their own food but depend upon the photosynthesis of plants. The chitinozoans are heterotrophs that exist between the multicellular animals and the autotrophic algae.

4
The Age of Procaryotes

The earth had traveled around the sun over one billion times before even the simplest form of living thing appeared that has left any trace of its existence. It was a world much different from today's. The barren, rocky surface that thrust above the primordial ocean was the dark gray and black face of volcanoes; the atmosphere had no oxygen and consisted of nitrogen, some hydrogen, carbon monoxide, and carbon dioxide. The carbon dioxide was less than 1 percent, but still as much as ten times the present concentration. And the oceans were shallow basins of hot wash of the basaltic surface.

The sediments were the detritus from the erosion of predominantly basaltic rocks of volcanic origin being deposited in an anaerobic marine environment. The reducing conditions of the atmosphere and oceans resulted in an appreciable amount of ferrous and sulfide ions in the seawater. The oceans were not to begin taking on modern characteristics until the recycling of sediments predominated over the erosion of basaltic rocks and the atmosphere contained free oxygen. This time was still far in the future.

It was a strange setting, quite alien to what we normally regard as conducive to life. But there it happened. Sometime before 3.4 billion years ago an aggregation of relatively small and simple organic compounds assembled in a lipid envelope and began to mediate elementary reactions. Those cells that were able to assimilate available substances and condense them into polynucleotides and polypeptides replicated and took on the nature of primitive bacteria. Metabolism developed from reactions the cells were able to use to degrade materials for their chemical energy and reactive products. In order to accomplish this, chemical energy had to be

25

held in a carrier and transferred to other molecules to be converted into activated derivatives. Pyrophosphates, particularly adenosine triphosphate (ATP), were adopted early and probably initially. Once activated derivatives were formed, they were energetically favored to follow spontaneous degradation—reactions that proceed on their own, but often slowly unless promoted by a catalyst.

This level of development probably soon gave rise to another organism that was able to draw from the inexhaustible supply of carbon dioxide for its carbon. In order to do so it needed a source of energy and it needed hydrogen. Both were in plentiful supply. All that was needed were chemical procedures which could be used to extract and harness them. The earth's surface was bathed in boundless radiation from the sun each day, and the visible light was absorbed by colored substances and converted to heat. If, instead of being squandered as thermal energy, the light energy were trapped and held in a chemical structure long enough to be used to generate ATP and reduce carbon dioxide, the organism would have a supply of energy even when no food substances existed.

It apparently was at this stage when the first ferredoxin appeared. Formed from a complex of polypeptide and the abundant iron sulfide, this biochemical became incorporated in the biochemical system early and has remained as a universal constituent of living cells ever since. As a component of the photosynthetic apparatus, ferredoxin was able to accept the energy of light absorbed by pigments and retain it at an electron energy level until it could be used to reduce carbon dioxide.

The earliest photosynthetic organism would then have required a pigment, ferredoxin, and a source of hydrogen. The donor of hydrogen for the beginning organism would have been substances which required the least amount of energy to extract the hydrogen and were still readily available. This supply appears to have been accessible organic matter.

Presumably, the earliest organisms were heterotrophs that thrived by metabolizing a reservoir of organic matter, but developed a rudimentary form of photosynthesis when the food supply neared exhaustion. There survives today a pigmented bacterium called *Athiorhodacea* that seems to be of this early stage of life. The *Athiorhodacea* is able to grow anaerobically as a heterotroph in solutions containing butyric acid and other organic nutrients by using the chemical energy it derives from them. But this organism is also able to absorb light and carry out the photocatalytic transfer of hydrogen and reduce carbon dioxide. Its source of hydrogen in this case is the nonutilizable organic matter.

But when the organic matter was no longer available or was in short supply, another source of hydrogen was needed. Photosynthetic bacteria evolved that were capable of using hydrogen sulfide as their hydrogen donor. This type of anaerobe is still extant today as the purple sulfur bacterium (*Chromatium*) and the green sulfur bacterium (*Chlorobium*) that are found in shallow lagoons and sea inlets where hydrogen sulfide is in abundance. In each of these types of photosynthesis, oxygen is not a by-product.

These bacterial forms of photosynthetic life may have been the dominant organisms on earth for several hundred million years. Certainly the fixation of carbon dioxide was being carried out for a considerably long time before organisms evolved to the level of being capable of oxygen-liberating photosynthesis. Eventually, however, because of the failing supply of the other hydrogen donors, or perhaps simply because of the sheer abundance, an organism became biochemically sophisticated enough to extract hydrogen from the most plentiful supply on earth—water. It requires ten times as much energy to remove hydrogen from water as from hydrogen sulfide, but the supply of water was inexhaustible. The organism that developed the photocatalytic breakdown of water became the cyanobacteria and began the 3-billion-year history of oxygen production.

It has not been established exactly at what time the blue-green algae emerged on the scene. The carbon isotope studies of Oehler[1] indicate that fixation of carbon dioxide was coterminous with the oldest microfossils 3.4 billion years ago. This process could have taken millions of years to develop and would still have coincided with the earliest fossils. Bulawayan stromatolites of Zimbabwe apparently are the result of cyanobacterial activity. Since they are dated at 2.6 billion years, an interval of 800 million years—longer than it has taken man to evolve from the level of a single-celled protozoan—passed from the first appearance of life to the time when the blue-green algae were growing in abundance in the tepid waters of the Archean seas.

There is no question that the cyanobacteria were among the ancient forms of life to appear on earth—and one of the most successful. These simple microscopic organisms thrive even today and occur in abundance in small freshwater bodies— ponds, ditches, and shallow lakes—during and immediately after periods of high air temperature. To some extent they are found everywhere, from the polar regions throughout the temperate zones to the warm waters of the tropics. And except for some bacteria, they grow where no other organisms can—in the 80°C (176°F) waters of hot springs of New Zealand and the Yellowstone.[2]

The bacteria and cyanobacteria belong to a particular division of life called the *Monera*. All other living things are either eucaryotic unicellular microorganisms or eucaryotic multicellular forms of life. Unlike the traditional two-kingdom system of plants and animals, evolutionary relations are better represented by a five-kingdom classification. Members of the *Monera* are the procaryotes, distinguished by the simplicity of their cellular structure; whereas all other forms of life are either singular or multiple eucaryotic cells in which the nucleus, the mitochondria, and other subcellular components are sheathed in membranes. In the procaryotes, the only membrane-bound object is the cell itself.

The cells of the cyanobacteria are encased in an outer mucilaginous sheath, a middle pectin layer, and an inner wall of cellulose. In contrast to the cell walls of the other algae, those of the blue-green algae contain amino acids, as do bacterial walls. The blue-green algae developed early as a sturdy but efficient packet of photosynthesis. By extracting carbon dioxide and using water as the hydrogen

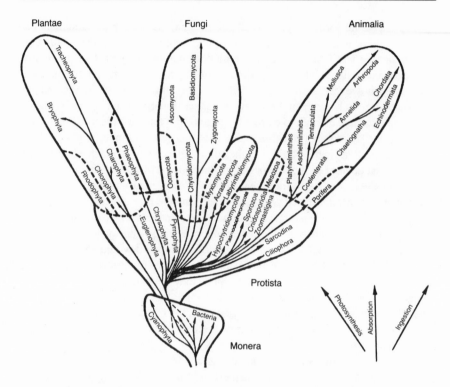

Figure 4.1. The five-kingdom-classification of life is based on three levels of development: the procaryotic (*Monera*), eucaryotic unicellular (Protista), and eucaryotic multicellular. Each level diverges in relation to modes of nutrition. *Monera* have photosynthetic and absorptive; the two higher levels are divided into photosynthetic, absorptive, and ingestive.

donor, they convert it to cyanophycean starch. These algae were and still remain the simplest food-producing plants.

The Archean ended 2.5 billion years ago with the climactic uplifting of mountains. For the Canadian Shield, this became the Kenoran province that added to the margins of the growing continent, extending from the Slave province in the northwest to Labrador eastward across Greenland to terminate in the continental shelves of the Atlantic between Greenland and Scotland. Comparable developments occurred in western Australia, in southern India, and in central and southern Africa. The sedimentary basins that were forming were of cratonal origin, in contrast to the volcanic sediments of earlier times.

When life began on earth, it was in an anoxygenic environment. The atmosphere contained a small amount of hydrogen, and the rocks and most minerals in solution in the oceans were in their reduced and lowest valence state. This is particularly notable for iron and sulfur. Ferrous salts are relatively soluble, but

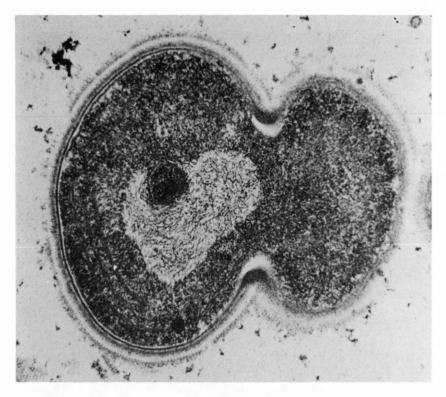

Figure 4.2. Cyanobacteria (*Gloeocapsa*) magnified 40,500 times in the electron microscope. This is a longitudinal section of a cell dividing. The light central area contains the DNA, the concentric striated lines are the lamellar where photosynthesis occurs.

oxidized iron is not. As a result, the concentration of dissolved iron in seawater today is extremely low (less than 10^{-7} molar), but this was not the case during the Archean. Sulfur, which is easily oxidized, exists principally as the sulfate ion in the present oceans. But in early times, it was as the sulfide from the dissolved hydrogen sulfide of volcanic origin.

The primordial life that evolved under these circumstances were the anaerobic microorganisms. They metabolized their carbohydrates by fermentative degradation, extracting the energy stored in the chemical bonds when carbon dioxide was reduced by photosynthesis to sugars, in the manner of yeasts today. It was a perfectly adequate biochemical method for a simple form of life under the environmental circumstances.

But the liberation of oxygen by the cyanobacteria in their photocatalysis of water introduced a perilous form of pollution to all life. Free oxygen is an extremely reactive agent that readily oxidizes reduced substances, mineral or biological. To avoid being destroyed by their own waste, the blue-green algae had to have the oxygen neutralized or removed from their immediate environment as it was generated. In the Archean oceans the most available and reactive chemical species for this role was the ferrous iron, and as the ferrous salts reacted with oxygen, the insoluble ferrous oxide precipitated. In time the evolution of oxygen was to change the character of the oceans. The blue-green algae were widespread and flourishing by the end of the Archean; after another 500 million years, they were generating so much oxygen that they were depleting the oceans of ferrous iron. Vast deposits of this precipitated iron banded chert from precipitated silica became the iron ore of North America around the Great Lakes and in Labrador, in the Hemersley region of western Australia, and in Mauritania of northwest Africa.

The climatic temperature was slowly declining from the 70°C (160°F) in the Middle Archean 3 billion years ago. Between 2.2 and 2.0 billion years ago the earth entered one of its cooling cycles and the first known ice age occurred. From that time on this was to be a recurring episode in the earth's history.

The last great deposition of banded iron formation was 2.0 to 1.8 billion years ago. Thereafter, with much or most of the iron oxidized, oxygen was leaked to the atmosphere. Some ironstone deposits from this time contain oölites and other structures indicative of shallow water deposition, showing that they were among the oldest shorelines in the earth's past. The remains of these can be seen today in Labrador and in Karelia, near the border of Finland and the USSR.

By this time the kinds of microscopic life forms had become much more diverse. This is the time of the Gunflint Formation, which left the multitude of bacterial and algalike microfossils. Nevertheless, all life was on the microbial level and confined to the oceans. The land stood sterile and barren of even the simplest form of vegetation.

Life in the oceans, on the other hand, was teeming. The efficient biological system of the blue-green algae for drawing on the unlimited reservoirs of carbon dioxide, water, and sunlight to construct their organic components allowed them to expand in an oceanwide ecological niche. By the Middle Proterozoic, they were so much the dominant life form that the time could be called the Cyanophycean period. And although the blue-green algae themselves are microscopic, they were leaving in shallow water monumental evidence of their presence.

Living at a time when there was no oxygen in the atmosphere, the early biological systems were without the benefit of an ozone layer to screen out the deadly ultraviolet rays. Any organism under these circumstances survived only in sheltered niches or at depths that afforded protection from the radiation. For this reason, cyanobacteria evolved from predecessors most suitably adapted for weak illumination. Eventually these phytoplankton evolved thick individual gelatinous sheaths, either colored or colorless, as protection against light too intense. Many lived in colonies or shapeless masses or layers of mucus, developed as the result of the dissolving and coalescing of individual sheaths. Being photoactic, they moved through any covering sediment that shut out the light completely, and as they did so their mucus cemented the particles in place. Many had sticky mucus sheaths that formed loose networks of filaments which sediments fell into, creating columnar shapes. So despite being perishable cellular matter, blue-green algae left enduring monuments in the widespread Precambrian stromatolites.

Around 1.5 billion years ago, the oldest deposits of calcium sulfate were being formed. This indicated that, although sulfate ions may have existed in solution for some time before this date, there was sufficient free oxygen at this time to oxidize the sulfur acids in the oceans. With the passing of nearly three-quarters of geologic time, the oceans must have been as large at this time as they are today, and they were taking on modern characteristics.

Having developed and evolved in an oxygen-free environment, all the microbiota that populated the oceans were anaerobic organisms, which is to say, they did not use oxidative respiration. They neither needed nor desired oxygen, which to them was a deadly poison. But with the last of the ferrous iron laid down in banded iron formations, from about 1.8 billion years ago onward free oxygen had been entering the atmosphere from the photosynthesis reaction of the blue-green algae. It was taking over a billion years for the oxygen from photosynthesis to oxidize all the substances of the earth's atmosphere and hydrosphere, but the pace went unabated. Eventually, this changing of the earth's environment from reducing to oxidizing conditions constituted an encroaching peril to the occupants who had been the undisputed masters of the earth for nearly 2 billion years.

Many that lived met the threat by developing oxygen-mediating enzymes, the oxidases, to protect their biochemical constituents from destruction by oxygen. Some presumably survived by retreating to oxygen-deficient niches and exist today as anaerobic bacteria. But the age of the procaryotes was coming to an end. Their dominance was foredoomed by the ever-increasing oxygenation of the air and water.

It was Louis Pasteur who discovered that obligate anaerobes cannot tolerate oxygen concentrations above 1 percent of the present atmospheric level. Following the oxidation of the salts and minerals of the oceans and the reduced gases of the atmosphere, the concentration of free oxygen in the air began moving toward the Pasteur point. The level of oxygen may have reached a critical level for the anaerobes by 1.4 billion years ago, when the changing environment brought about the emergence of the eucaryotes.

5

The Advance of the Eucaryotes

Only the very simplest type of cell could have lived at the high temperatures of Archean times on earth. But by 1.3 billion years ago, climatic temperatures had declined from around 70°C (160°F) to 52°C (126°F)—a temperature just below the spontaneous denaturation temperature of complex proteins.[1] The eucaryotic cell developed when the overall temperature had dropped to a level compatible with its complex molecular and cellular structure. It was also better equipped to cope with the presence of free oxygen than the anaerobic organisms.

From the emergence of the eucaryotes between 1.4 and 1.2 billion years ago onward, they began to advance the efficiency of their revolutionary new cellular organization. But as long as the atmospheric oxygen remained less than 1 percent of today's level, only the fermentative metabolism of carbohydrates was possible, and they were confined to the microscopic level of yeast.

After 500 million years the earth's environment had undergone an appreciable transition, and the eucaryotes were reflecting the change as their complex cellular structure continued to evolve into more efficient composition. By 750 million years ago they reached the level of complexity of chitinozoans.

The climate was cooling and the oceans were much as they are today. The only forms of life were still confined to the seas, and the continents, now approximately 30 percent of the earth's surface, stood broad and barren, with rivers and lakes—but no plants or animals. The day had extended to 20 hours with 442 days in the year,[2] and the moon was no longer the awesome sphere dominating the night sky, for it had been receding from the earth throughout the ages, until toward the close of the Precambrian it appeared only about twice its present size.

Around this time some eucaryotic cells began to forsake their solitary ways and began to share a colony as a loosely associated collection of individuals. In a comparatively short time the colony took on a character of its own as the individual cells became more dependent on being a part of it. The cells interacted with each other by excreting chemicals and ions which affected the biochemical synthesis governing reproduction and products in each other. In this way cells within the colony became specialized and different from each other in the group. The colonies that were most successful with this group-living in the gathering of nutrients and protection against predation evolved and passed on the genetic propensity for differentiation necessary for colonial organization.

The scenario is not purely conjectural. There are found today living systems in this intermediate stage of organization in which cells become colonies, united and specialized, but not to the extent that they may be classified as multicellular organisms. Among the green algae there is a one-cellular form which has a chloroplast, an eyespot, and two flagella, or threadlike projections, used for locomotion and the movement of water currents. Within this family are the *Pandorina,* some of whom form colonies of four to thirty-two cells. These colonies are not merely aggregates, because the cells swim by coordinated movement of their flagella. The *Gonium* is another member of this group that forms colonies.

But the most evolved colony formation is carried on by the *Volvox.* This genus of pale green flagellates forms a colony of 500 to 50,000 cells arranged in a hollow sphere about one-fiftieth of an inch in diameter. Whereas the cells of the *Pandorina* and *Gonium* look alike, the cells of the *Volvox* become specialized; the cells at the front of the ball have larger eyespots, and only some of the cells reproduce themselves. Thin strands of protoplasm connect the cells to one another in the colony. The colony is reproduced when cells in the back begin to divide, producing a new small ball of cells that is released to the inside of the parent colony. When the old colony dies, the young colonies inside are released to disperse and repeat the cycle.[3]

From colony formation the eucaryotes were crossing the threshold to become *Metazoa,* or multicellular animals. As there are of most stages of evolutionary development, there are living species which have resisted change and have survived in the ecological niche that was prevalent at the time that level of development was widespread. The sponges are animals that lie at the borderline between the unicellular and multicellular development.

The surface of the sponge is formed by the ectosome, a layer or layers of cells perforated by minute pores. The body is crossed by numerous canals which run into and out of flagellated chambers like little thimbles lined with cells bearing a funnel-shaped collar. The sponge takes in food and oxygen by drawing water through the pores and canals to the flagellated chambers with the collar cells and out again through vents, the waste products from digestion and respiration being discharged by the exhalent current.

On the East Coast from Nova Scotia to South Carolina there is a small sponge found growing commonly on oysters called *Microciona prolifera*. It begins life as a thin crustation, but as it ages it sends up vertical, fingerlike lobes with rather large openings at the ends as vents. As the sponge gets older, these projections increase in number to form a bushlike structure with interwoven branches. Some specimens grow as tall as eight inches. The color of *M. prolifera* varies seasonally, but in the summer and autumn it is tomato red.

In 1907, H. V. Wilson,[4] a biologist with the University of North Carolina at Chapel Hill, spent his summer in the U.S. Bureau of Fisheries laboratory at Beaufort Harbor doing research on his special interest, the degeneration and regeneration of tissue. Being aware that *M. prolifera* had remarkable regenerative powers, Wilson decided to experiment with the little red sponge.

He cut the sponge into small pieces with scissors and strained the tissue through fine bolting cloth. Holding the pieces of sponge in the cloth as in a bag, he immersed the cloth in a saucer of filtered seawater and squeezed it with forceps. Red clouds of cells passed out through the cloth into the water and quickly settled to the bottom of the saucer like a fine sediment. After the sponge was disintegrated and the constituent cells strained into the saucer of water, Wilson carefully studied the behavior of the sediment.

As he watched, the cells began to fuse with each other. After a short time, they had conglomerated into many little balls. Protrusions of protoplasm or pseudopodia then reached out from the balls over the surface of the saucer. When they made contact with other groups of cells, the conglomerates came together to make one. Eventually, all the balls of cells fused into a single incrustation. Differentiation of the cells began, and as the days went by, flagellated chambers appeared in great numbers, connected by canals formed in the structure. Short vent tubes began to grow vertically from the incrustation. After six or seven days, the little red sponge had regenerated itself.

Sponges are at the lowest level of development of multicellular animals. But the experiment was remarkable in showing that metazoa result from individual living cells which have come together in coordination in a specific way to create a higher order of life. Moreover, it supported the thesis that the single cell is the primary unit of developmental function.

Wilson tried next specimens of the coelenterates, that phylum best known by the jellyfish, which is at a transitional stage that has led to all higher animal phyla. The structure of coelenterates, being composed of two layers of tissue, ectoderm and endoderm, is more evolved than that of sponges. Nevertheless, when polyps of hydroids were cut into fragments and squeezed through fine cloth, these dissociated cells also selectively reaggregated themselves to reform the animal.

Little came of Wilson's fascinating experiment for the next thirty years. Then in 1939, Johannes Holtfreter,[5] a biologist at the Kaiser-Wilhelm Institute in Berlin, dissociated the cells of a frog embryo in a solution free of calcium and magnesium ions. When the suspension of cells was allowed to stand undisturbed, the cells started to sort themselves, and eventually they reformed the original embryo. The experiment was performed with avian embryos in 1952 by A. and H. Moscona,[6] followed by other researchers working successfully with embryonic tissue from mammals. For reasons that will be explained later, the experiment works only for the early stages of the embryo.

The human body, as well as that of any other metazoan, develops from one cell, a cell that is replicated repeatedly, first forming tissues by making layers of a definite pattern, these tissues then growing into organs. In this way the embryo reenacts the evolutionary stages of development that took place hundreds of millions of years ago. In the end the fully structured animal with its own individual qualities is built of trillions of cells working together as a unit. It has been estimated that the number of cells in the human adult is of the order of 10 trillion, and all of these are ultimately derived from the single fertilized ovum. It seems like a very large number, but only 41 generations of cell division would be needed to attain it.

Multicellular differentiation meant a big increase in the number of genes to express the function of not only the single cell, but of the whole organism. The functions of the higher animal result from differentiation of cells: that is, cells specializing to generate various tissues, organs, bone, and hair. Because of the manner in which metazoa evolved from organized amassing of progeny of a single cell, every cell in the body has the full complement of genes and chromosomes for the complete animal. They differ in that not all the genes are expressed. Only those genes that are for producing proteins relevant to the specific cell's function are used; all other genetic material in the particular cell is shut off from transcription by basic proteins called histones that combine with the gene and prevent it from being copied. In this way eucaryotic cells were able to achieve and coordinate various functions from basically the same cell and were able to advance from the unicellular stage to more complex forms of life.

About a hundred million years was needed to advance from the tissue level of development of coelenterates to animals with organs, but when the breakthrough was made, the effect was explosive. The new type of organization of tissue for interacting with the environment was more flexible than its fairly passive predecessors, and this type of animal life quickly spread throughout the world and became dominant. These animals—the arthropods, mollusks, and others— grew exoskeletons of chitinlike material or mineralized shells to protect their fragile organs and tissue, but they remained in their sheltering ocean depths, and the land was still barren.

While the eucaryotes were advancing, one of the great mountain systems in the earth's history—the Appalachian-Caledonian—began to be thrust up. Between 1,100 and 750 million years ago, long periods of sedimentation had laid down platforms on the margins of North America and Europe. But the ancient ocean that separated the two land masses shrank as powerful forces within the earth brought the continents toward each other. When eventually they collided, the sedimentary platforms caught in the collision buckled and folded, pushing up towering peaks. From the heat and pressure, limestone was metamorphosed to marble, shales to slate and schist, sandstones to quartzite, and intrusions of magma formed bodies of granite. Whereas the mountains of five or six episodes more ancient than the Appalachian-Caledonian have long been obliterated by erosion, the Appalachian range in eastern North America and the Caledonia mountains of Scotland, worn and sculptured by the ceaseless action of water, stand today among the oldest mountains on earth.

The Late Proterozoic, which began as warm and moist, ended cold and dry as glaciation extended over eastern Canada and the earth experienced another ice age. With the start of the Cambrian, the climate grew mild. The climatic temperature was around 34°C (111°F), and atmospheric oxygen was now probably several percent above the Pasteur point. Trilobites, looking like a type of horseshoe crab, inarticulate brachiopods, primitive mollusks, the echinodermates, ancestors of the starfish, annelids (segmented worms), and the arthropods, primitive centipedes with segmented bodies, crawled along the sea bottoms.

The Cambrian, lasting from 570 to 500 million years ago, was followed by the Ordovician, with a major expansion of the invertebrates into new species, genera, and families. It was an age of marine animals with mineralized skeletons and shells, corals, starfish, brittle stars, and crinoids that colonized the shallow sea floor. The cephalopods, some straight, some curved, and some coiled shells, appeared, and among them was the nautilus, much like today, a formidable predator cruising in the ocean depths.

By the time of the Ordovician period 500 million years ago, with shallow seas covering large areas of the world, simple plants began to adapt themselves to life on the fringes of the land. It was a new condition in which they had to develop structural support against gravity and a vascular system to carry their fluids upward to exposed parts. Then around 450 million years ago there appeared in fresh water in what is now Colorado an animal that was probably a jawless fish. Its fossils are so rare and fragmentary that the form of the animal is unknown. They are found in the Harding sandstone as dissociated and broken plates of bony structure, but they indicate the presence of the missing phylum, the chordate with vertebra, our ancestor.

Since most fossil records are of marine origin, the account of freshwater and terrestrial development is more difficult to assemble. But colonization of the continents apparently occurred during the Silurian (420–395 m.y.) and Devonian periods (395–345 m.y.). Probably it began with the simple plants, followed by arachnids, scorpions, and millipedes. North America was low and flat during this time, except for mountains and volcanoes in the eastern United States and Canada. Europe was mountainous with arid basins. It was around 370 million years ago that an amphibian ancestor of all terrestrial vertebrates, including man, pulled himself out of his freshwater habitat onto a riverbank and found himself in Greenland. Or at least that is where his fossil remains were found 370 million years later.

Colonization of the land areas by plants led to the Carboniferous age from 350 to 270 million years ago. Large trees, some 100 feet tall, stood in swamp forests throughout many continents. There were relatives of the horsetails, the club mosses, and tree ferns of today, and there were primitive gymnosperms of conifers. Insects were diverse and included a giant dragonfly with wings 28 inches long, a 5-foot millipede, and enormous cockroaches and scorpions. Trilobites were nearing the end of their long history, and among the vertebrates were large sharks. Amphibians, descendants of the air-breathing vertebrates of the Devonian, were present, and the synapsids, the earliest known reptiles.

During the Permian period that followed, 280 to 225 million years ago, the synapsids were primarily predacious and they became dominant. A subclass of the *Synapsidia* were the therapsids, small, unimpressive relations, but active carnivores. During the Triassic period (220–180 m.y.) these small creatures developed special secreting glands to nourish their young with milk and became the first mammals.

The continents, like huge rafts of granite slowly skimming on the earth's mantle, had become fused within the preceding hundred or more million years into a supercontinent called Pangaea. But during the Triassic, the parts that were eastern North America and New Zealand were experiencing an outbreak of volcanic eruptions. Africa and South America as a unit swung away from North America, dividing Pangaea into Laurasia of the north and Gondwanaland to the south. India and Antarctica split from Africa.

As the Triassic entered the Jurassic period, the great trees of the coal swamps had already yielded to ginkgos, conifers, and cycads. Dinosaurs filled nearly every niche of terrestrial life, many attaining gigantic sizes. And a rift formed between North America and Europe. It was a small narrow body of water much like the Red Sea or the Gulf of California, but it was the beginning of the Atlantic Ocean. Mountains arose from Alaska to Mexico, and there were eruptions in the northwest.

The Cretaceous period saw the appearance of flowering plants. Their propagation by airborne seeds had an explosive effect, resulting in their being spread rapidly throughout the world. Bees followed in their wake. The mammals, living in the shadow of the dinosaurs for 150 million years, were on the sidelines as the giants lumbered into extinction.

With the decline of the dinosaurs 70 million years ago, the mammals rose to become the dominant form of animal. The concentrated energy in the seeds of flowering plants made possible the rapid metabolism for flight, and birds evolved. In western United States, the Laramide orogeny, one of a succession of mountain-building events from the late Devonian, was climaxing by plutonism that consolidated the marginal belt to the continental crust.

Sixty-five million years ago there were small forest-living creatures with the build and gait of squirrels. They were not yet specialized in tree climbing, but these were the earliest primates, a separate order of *Mammalia*. During the Oligocene 40 to 25 million years ago, the Alps and Himalayas were thrust up. Saber-toothed tigers evolved to prey upon the grazing mammals.

The Miocene, 25 to 10 million years ago, saw the uplift of the Sierras and the Rockies. Except for rodents, all recognized mammal families already existed. The fauna included deer, hyenas, the earliest giraffes, and the bovines. A huge bear-dog, Amphicyon, was conspicuous, and mastodons spread into North American grasslands to join the evolving horses and camels as grazing became dominant over browsing life.

When the Pliocene began 10 million years ago, a chill was on the tropics and grasslands were widespread. The mammals reached the peak of their size and abundance. It was about the middle of the Pliocene, 4 or 5 million years ago, when a scavenger appeared on the plains of eastern Africa, competing with the hyenas for what food he could snatch from kills by the predators much stronger than he. He was an errant primate that remained on the ground while the others developed a way of life in the trees. He lacked the teeth and the speed of a carnivore, but his cunning and group actions gave him an advantage that allowed him to survive.

Eventually he began to shape stones as weapons and tools. His teeth show that he evolved on a diet of roots, fruit, nuts, and other edibles he could gather. And his bad digestion—he couldn't eat green fruit and some raw foods like the baboon—led him to use fire to predigest foods he couldn't eat raw in order to expand his available food supply.

This of course was man. He evolved as a separate species sometime around 4 and 5 million years ago. Life had existed for 3.4 billion years; for 2.7 billion years, or 80 percent of the time, it remained at the level of one-celled organisms. Only within the last 2 percent of the time that life has existed on earth did mammals appear. And since the appearance of life in the Fig Tree and Onverwacht Series in South Africa, 99.9 percent of life's existence passed. Only in the last 0.1 percent of the time span did man finally evolve.

Why did it take so long?

What happened over a billion years ago, when the temperature was becoming cooler, the oceans were being oxidized, and free oxygen began to accumulate in the atmosphere, that could have created a cellular life form so advanced over the procaryotes? Why did becoming multicellular produce so explosive an innovation? And why was life confined for nearly 90 percent of its existence to the sea, venturing out on the land only a few hundred million years ago?

Our consciousness and feeling of free action are so far removed from the properties of terrestrial materials that it challenges the imagination to bridge the two natures. But we are looking at the products of an astronomical number of selective processes that carried evolution through nearly 4 billion years. When we look at the organization of our biological composition we find that the road to our being was not an even climb, but one that rose by stages like tiers of a step pyramid. At the base of our multicellular structure lies another stage of development that evolved its own level of sophistication and complexity.

6
Life's Cellular Nature

Throughout all but the last moments of his existence as a species man has lived in a world bound by the limits of his senses. The gradient of sizes shrank toward the small until no longer perceptible, and loomed into the large until it engulfed the horizon. The biological realm existed in great variety, each species reproducing its own that grew in size from birth to adulthood. It appeared that within a species there was no physical component smaller than the organism itself.

Some creatures, such as small insects, did seem to appear suddenly as though they had crossed over from another dimension. With no observable origin, spontaneous generation for the elusory small was easier to imagine than was an entire level of existence below the visible in a drop of water. Imagine then the sense of discovery in being the first to pass the limits of our world into another, into regions populated by bizarre and alien beings. Thus was the experience of Anton van Leeuwenhoek.

Leeuwenhoek was a draper in Holland who had no formal scientific schooling but a deeply absorbing hobby of grinding lenses. The microscope has been invented in 1590 by the Dutch brothers Francis and Zachary Janssen when they placed a lens at each end of a tube and obtained a magnification of ten times. But it was a crude device and hardly more advanced than the magnifying glass. Leeuwenhoek, on the other hand, made high quality lenses of a very short focal length with a magnification of 50 to 275 times, a resolving power far exceeding the early compound microscope. He ground lenses of many sizes—some no larger than the head of a pin—and mounted them between two thin brass plates. And as Leeuwenhoek peered through his lenses at samples of rainwater, pond water, and scrapings from his own teeth, he saw a strange, tiny universe teeming with life.

For the first time in history the "very little animalcules," as he called them, describing bacteria and protozoa, entered our world of awareness. Leeuwenhoek was a man of insatiable curiosity and powers of careful observation. He discovered a new world and he explored it with every opportunity. He saw things that never before had been seen. In long, rambling letters to the Royal Society of London he explained that weevils in granaries were not bred from wheat, but were grubs from tiny eggs; that fleas bred in the regular manner of insects and did not sprout from sand or dust; that shellfish were spawned, not generated out of sand; and that embryos of fresh water mussels were sometimes eaten by his little animalcules. He studied the circulation of the blood in the capillaries of the tail of an eel, the web foot of a frog, and ear of a rabbit; and in his letters he illustrated the red corpuscles of various animals. In 1677, he described for the first time the spermatozoa of insects, dogs, and man. For fifty years he explored and described to the world the fascinating universe he had discovered on the other side of his lenses.

While Leeuwenhoek studied his tiny universe, Robert Hooke, using a crude microscope with lenses of fused glass, observed the honeycomb appearance of thin slices of cork and referred to the spaces as cells. He had seen the cell walls, and for a hundred and fifty years these features were believed to be the basic structure of plants. It was the German botanist Hugo Mohl who realized that the basis of life was not the cell wall seen by Hooke, but the translucent, semifluid substance inside. He named it protoplasm.

Animal cells, lacking the cell wall of plant cells, are smaller and, except for a few like eggs, spermatozoa, and blood corpuscles, were more difficult to see with the early microscopes. Not until Lister invented the achromatic microscope in 1827 was it possible to make a detailed study of cells. Four years later Robert Brown detected a dark speck in the protoplasm of plant cells and discovered the nucleus. Matthias Schleiden in Germany recognized that the nucleus was associated with cell division and realized that this was the manner in which cells come from preexisting cells.

A year later, following a dinner conversation with Schleiden, the German anatomist Theodor Schwann sensed that the animal cells that he was studying behaved like Schleiden's plant cells. It occurred to him that the cells were like organisms, performing all the necessary activities of single-celled microbes. In plants and animals they are organized in a concerted union to form a whole organism on a higher level of existence. Then suddenly, in one of the most sweeping generalizations in the history of science, Schwann saw a great common denominator for all living things—plants and animals were constructed from the same basic unit: the cell. Schwann published his thoughts in 1839 in a paper entitled *On the Correspondence in the Structure and Growth of Plants and Animals,* and the cellular theory of life was established.

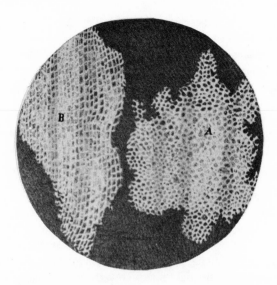

Figure 6.1. Drawings of a piece of cork from Hooke's book
Micrographia. A is a cross section; B is a longitudinal section.

The cell is the smallest autonomous biological system—the basis of life—and
all living things consist of either a single cell or a combination of cells. The cell
is, in effect, a biochemical machine capable of self-replication. Given a source
of energy and the essential raw materials, it can construct its own components
and reproduce its entire composition. Considering the fact that the number of
individual constituents can run into tens of thousands packed in a sphere so small
that it needs a magnification of several hundred times to be visible, it is a remarkable package.

When viewed through a conventional light microscope, all cells appear to be
capsules of transparent and largely homogeneous material. The only structures
discernible are the chromogenic particles such as the chloroplasts of the plant
cells, which contain chlorophyll, or pigmented substances in animal cells. In
order to study cells the microscopist uses various chemical stains to color selectively or darken internal features of the cell. For over a century this was the
principal means to observe cellular composition and activity.

When the biologists of the nineteenth century began viewing the microscopic
world, they found it populated with such a variety of unicellular organisms that
it strained their method of classification. There were protozoa, single-celled animals with relatively large and complex cells; there were bacteria, smaller and
simpler; and there were the algae and fungi. It was an odd assortment of creatures
on a level where the distinction between animal and vegetable became blurred.
But of all the unicellular organisms, only the bacteria and the blue-green algae
lacked a distinct nucleus, and consequently were grouped together in a class that

came to be known as the procaryotes; the cells of all other plants and animals had membrane-bound nuclei and were called eucaryotic cells.

The light microscope went through several modifications and innovations over the century to improve its capability. There were the phase-contrast, the inference, the polarizing, the ultraviolet, and the dark-field microscopes; but detailed descriptions of cellular structure always remained agonizingly beyond the observable. Resolution, dependent on the wavelength of light, was limited to about 0.24 micrometer.* Since procaryotes range from 0.3 to 2 micrometers in diameter, and eucaryotic cells with their more complex composition about ten times larger, the resolution was able to show the cell, but was not fine enough for structural detail.

Then in 1924 Louis de Broglie hypothesized that electrons should travel in characteristic waves like light with the wavelength being inversely proportional to the root of the voltage. Theoretically, electron beams could have wavelengths much shorter than light, and a microscope using electrons instead of light would have a much greater resolving power—a 100 kilovolt electron beam would improve the resolution over the light microscope by approximately 140,000 times. Based on this principle, the first electron microscope was built in Germany in 1932 using magnetic lenses, and the Siemens-Halske Company produced a commercial prototype in 1938.

The electron microscope revolutionized the knowledge of the way cells are organized. It showed that the difference between the two kinds of cells was more than the presence or absence of a nucleus—it was an extreme difference in subcellular architecture. After the complexity of the structure and chemistry of eucaryotic cells became known, it was apparent that the gulf between procaryotic and eucaryotic cells is so great that the development of the latter must have been one of the major steps in evolution.

Both kinds of cells are bound by membranes of similar profiles, resolvable at high magnification into triple-layered structures about 0.008 micrometer thick. In the procaryotic cells there are no membrane-bound units; the nuclear material and cytoplasm appear uniform and unseparated. The principal structural units are subcellular particles about 0.01 micrometer in diameter known as ribosomes where protein is synthesized.

The eucaryotic cell, on the other hand, as observed under the electron microscope, is no longer a homogeneous cytoplasm, but a labyrinth of sections separated by membranes. This system of membranes results from convolutions of the cellular membrane, making all membranes continuous with one another. The nucleus of the cell is surrounded by a membrane and contains nucleic acids and proteins. In other regions of the cell there are discrete organized structures called organelles, which play specific roles in the cellular functions: ribosomes for the

---

* 1 micrometer = 0.001 millimeter (0.00004 inch)

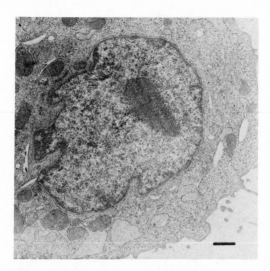

Figure 6.2. Cross section of the major part of a mammalian cell viewed through the electron microscope showing the nucleus, nucleus membrane and nucleus area. In the cytoplasm one can see vesicles and mitochondria. The bar in the lower right-hand corner represents 1,000 nm; magnification is 7,500 X.

synthesis of proteins; mitochondria as powerhouses of the cell where pyruvic acid is oxidized to carbon dioxide and water for energy to be used by the rest of the cell; Golgi bodies, lysosomes, and others.

The nucleus is the storage site for the blueprint to reproduce the entire organism. In eucaryotes it is a distinct organelle, whereas with procaryotes it lacks a nuclear membrane and floats freely in the cytoplasm. The hereditary information for the cell is stored in the chemical composition of a long linear polymer called deoxyribonucleic acid (DNA).

The DNA is the informational molecule for a living system. Its coded structure determines the amino acid sequence of proteins. But proteins create the organism; they are the workhorses that make it all possible. Proteins are responsible for the structure, for the transport of material, the regulation of processes, and above all, they are the enzymes that catalyze all biochemical reactions. All cells, procaryotic or eucaryotic, have to synthesize proteins, and this task is carried out in small subcellular particles called ribosomes.

Eucaryotic ribosomes are about 0.025 micrometers in diameter with a total molecular weight of 4,000,000 for higher organisms, about 2,700,000 for bacteria.[1] Each ribosome is composed of two subunits, one approximately twice the size of the other. The smaller subunit has one large molecule of ribonucleic acid (RNA); the larger has two RNAs of unequal size. Additionally, there are about

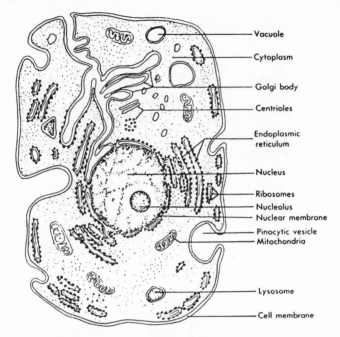

Figure 6.3. Diagram of a eucaryotic cell.

20 different proteins in the smaller unit, 40 in the larger. As far as can be determined, the proteins are present as only one molecule each.

The functional arrangement of the constituents of ribosomes is highly specific. This constancy of organization is manifested by the fact that at least some ribosomes have been crystallized. The ribosomes of avian embryos, for instance, can be induced to crystallize merely by cooling the cells.[2]

Unlike ribosomes, the mitochondria are common only to the eucaryotes. These are sausage-shaped particles 0.5 to 1.0 micrometer wide and 5 to 10 micrometers long. They are surrounded by an outside membrane (thylakoid) and are partitioned into a series of chambers by inner membranes called christae. Plant cells tend to have fewer mitochondria than animal cells. The number range from one for the unicellular alga *Microsterias* to 500,000 for the giant amoeba *Chaos chaos*.

When seen in a living cell, mitochondria appear to be in constant motion. They are the respiratory centers for the cell. In the mitochondria the carbohydrates, fats, and to a lesser extent, proteins, are metabolized to supply energy for the cell. In a series of chemical reactions catalyzed by enzymes, substances of higher chemical energy are broken down to compounds of a lower chemical level, and the energy difference is used to synthesize the high-energy molecule adenosine triphosphate (ATP). ATP is the energy-rich chemical that the mitochondria distribute around to the various sites of synthesis to activate compounds for conversion to other chemical structures.

In green plants the organelle that serves to trap sunlight is a structurally complex particle called the chloroplast. The number of chloroplasts for plant cells varies. In some algae, such as the filamentous *Spirogyria,* there is only a single chloroplast; whereas, a cell in the spongy part of a grass leaf may have 30 or 40 chloroplasts.

Other organelles in eucaryotic cells evolved as forms of life became more complex. The Golgi apparatus, for instance, is a system of membranes used in the packaging of proteins to be secreted, such as digestive enzymes. These proteins are accumulated in the Golgi apparatus, carbohydrate is added, and a large number of molecules are wrapped in a membrane. The package is then passed to the edge of the cell, where the contents are released to the outside.

Lysosomes are particles about the size of mitochondria, but without the highly organized structuring. These organelles are membranous sacs of digestive enzymes that can dismantle large proteins and nucleic acids. Their purpose had not been apparent until recently, but when realized, it revealed the extreme sophistication of the biological cell.

The molecular components of all living things are in a constant state of flux. Constituents are continually being broken down and replaced; even the highly organized structures like the mitochondria have a transient existence. The lifetime of mitochondria of the liver, for example, is 10 to 20 days. Why this should be the case was not immediately obvious.

The reason has to do with information storage. The structures of the cells' polymeric molecules and subcellular units are the result of an enormous number of information bits—and each configuration must be exact to be efficient. A system of repair for defects that occur would require another information system of nearly the same magnitude. It is far simpler merely to replace a defective protein or organelle with a newly synthesized one. Therefore, whenever the efficacy of a cellular component falters, the part is dismembered by the lysosome and its debris is fed into the furnace of an intact mitochondrion to fuel other syntheses.[3]

The lysosome has an additional function. When the cell dies, the lysosome sac bursts and the released enzymes digest the cellular biopolymers back to their monomeric units, thus obliterating the chemical organization that made the cell a biological entity—and returning the building blocks to be used again.

All cells are encased in a membrane structure with unique properties for the containment of cellular components and as a semipermeable barrier to the outside environment. One of the universal features of cellular membranes is their protein-lipid nature, and of the lipids, phospholipids have characteristics that make them particularly suitable for the membrane.

When eucaryotes evolved, they retained the architecture of the cellular membrane and extended it greatly. Viewed by an electron microscope the interior of a eucaryotic cell is seen as a maze of sections housing the cytoplasm and the subcellular particles. Invaginations of the membrane intrude deep into the cell and the elaborate cytoplasmic membrane system meanders through the cell as a continuation of the cell membrane. This involved system of membrane—the endoplasmic reticulum—serves more than to increase the surface area: it is now recognized as an important part of the cell for the manufacture of cellular products. Within the structure are enzymes which control reactions in strategic parts of the cell. The endoplasmic reticulum is a major adaptive modification of the eucaryotic cell not found in the procaryotes. Its channeling through the interior of the cell apparently permits some direct transport of various molecules and ions from one part of the cell to another, and even out of the cell.

The complexity of the eucaryotic cell is matched by its mode of reproduction, called mitosis. Whereas the DNA of procaryotes remains floating in the cytoplasm as a long strand closed in a loop, the DNA of eucaryotes coils into chromosomes that look like beads on a string, each chromosome containing a definite group of genes. The number of chromosomes varies with species: the fruit fly has 8; the onion, 16; man, 46; and cattle, 60.

During much of the lifetime of the eucaryotic cell the nucleus appears to be a homogeneous structure, except for a small discrete spherical area inside the organelle called the nucleolus. It would appear that during the resting stage, with the chromosomes diffused in aimless disarray, the nucleus is devoid of activity. This is the period, however, when the monomeric chemicals are being assembled to replicate the DNA molecule.

When the time to reproduce arrives, two pairs of minute structures—the centrioles—just outside the nucleus begin to move apart. As they do so they stretch between them gossamer threads called spindle fibers. The centrioles move toward opposing ends of the cell, the chromosomes commence to coil and condense.

The membrane of the nucleus then starts to crack and crumble, allowing the spindles to reach across the whole cell. By this time the chromosomes have become well-defined, rod-shaped structures, and the spindles attach to small bodies in them. The chromosomes are then pulled to the equator of the cell where they are lined up like sausages to be divided equally between the two daughter cells.

Once the division between the chromosomes is made, the spindles begin to reel them toward the poles of the cell. After this the chromosomes relax from their compacted whorls. Even as the chromosomes begin to uncoil and stretch, a nuclear membrane is spun around each group. The cellular membrane then comes together in the middle and pinches off, separating the original cell into two fully independent daughter cells.

It takes the human cell about 18 hours to regenerate, the bacterium about 18 minutes. But the bacterial DNA is like a single chromosome consisting of about 2,500 genes with approximately a million monomers strung together in a linear polymer. Fully extended it would be about 1.5 millimeters long. The DNA of a mammalian cell, on the other hand, can contain 100,000 genes and be a polymer of 4 billion monomers, reaching 50 inches if extended.

The procaryotes are the bacteria, the cyanobacteria, the recently discovered chloroxybacteria,[4] some multicellular organisms such as the actinobacteria, and the fruiting myxobacteria. The eucaryotes include most of the familiar organisms: the seaweeds, protozoa, fungi, plants, and animals. The procaryotes possess metabolic pathways for the fermentative breakdown of organic matter, the fixation of atmospheric carbon dioxide and nitrogen, the oxidation of hydrogen sulfide to sulfur, and the biosynthesis of fatty acids and isoprene derivatives, such as porphyrins. With the rise of eucaryotes came aerobic respiration and steroid synthesis. Both cellular types have photosynthetic members.

These pathways, along with the mechanisms for replication and synthesis of macromolecules, represent most of the basic cellular activities of all living things. Since they exist even in the microorganisms, we face the realization that the distinctions we draw between ourselves and other forms of life are not so biologically significant as we prefer to believe. That before there were plants and animals, before the continents contained life, before cells evolved the means of multicellularity, all the great biochemical innovations already existed.

7
Molecular Architecture

The human body is composed of approximately 10 trillion cells—each like a single-celled organism, taking in nutrients and maintaining its existence while performing a specific function in the creation of the whole person. And from the intricate orchestration of cellular functions come the consciousness, the sensations, and the life processes that allow us to exist in our dimensional world. At the basis of the numerical and behavioral complexity of our existence is the autonomous unit—the cell.

The fundamental unit of our composition is the eucaryotic cell, too small to be seen except by a magnification of several hundred times. But the volume and complexity of the eucaryote is a thousand times greater than that of the procaryote. Despite their extreme smallness, the procaryotes are not themselves elemental. For they too are autonomous units whose vitality and integrity stem from the orchestrations of constituents of another order that lies at their base.

The bacteria are the simplest form of life. Across a chasm into the depths of smallness beyond the reach of even the electron microscope stirs a world that bridges the animate and the inanimate. It was out of this world that grew the structural units that created living systems. Those structural units are chemical molecules.

The science of chemistry itself had to develop for 200 years before it was able to deal with the complex chemistry of biological systems. By the eighteenth century, scientists had established several basic principles of chemistry and physics which were valid for inorganic substances, but the complexity and less predictable nature of biological systems seemed to exclude them from these physical tenets. This impression was given formal recognition in 1707 when the

German physician Georg Ernst Stahl enunciated the concept that life was governed by special nonphysical laws. This was the theory of vitalism, which held that all living things contained a vital force, an inseparable, nonphysical component that directed and made possible the functioning of all life processes.

The early chemists were aware that substances from living things were clearly different from the mineral world and fell into distinct categories. There was the white, chalky starch that was extracted from wheat, potatoes, and rice and was used to stiffen the collars of burghers. There were oils from plants and animals for food and fuel in lamps. Being insoluble in water and greasy to the touch, fats and oils were lumped together as lipids. And there was a third group found in foods that differed from the other two. This was the albuminous substances in solution in the whites of eggs, in milk, and blood. Members of this category were particularly unstable and coagulated when heated.

Unlike most inorganic compounds, products of organisms were unstable to heat, usually changing irreversibly, and most could be burned. Elemental analysis showed starch to be composed of carbon, hydrogen, and oxygen; lipids were mostly carbon and hydrogen with a small amount of oxygen; and albuminous substances contained carbon, hydrogen, oxygen, nitrogen, and sometimes sulfur and phosphorus. With so much carbon and hydrogen, it became apparent why these substances burned. Except in limestone, carbon is not a common element among minerals. With living systems, it was always present.

The principles and methods of chemistry were established by the nineteenth century, and much of modern chemistry has grown from the work of Jöns Jacob Berzelius in Sweden. In the early 1800s Berzelius determined the exact elemental composition and formula of some 2,000 compounds. He also gave the world many chemical words, including *polymer* for a large molecule formed by the linking together of small subunits, and *catalyst* for a substance that promotes a chemical reaction without itself being consumed.

In Berzelius's time, a few pure biological compounds were already commonly known. Several crystalline sugars had been isolated from a number of sources and urea was known as crystals found when urine evaporated to dryness. The purified compounds derived from living organisms seemed to have many common properties which they did not share with inorganic chemicals. Berzelius coined the word *organic* and divided chemistry into inorganic and organic, the latter being the domain of compounds from living systems, distinct and separate from inorganic chemistry.

The chemical distinction was made. The Vitalists were satisfied. No matter how life originated, it remained a philosophical question because its worldly substance could not have come from the soil or from minerals. If there were to be a scientific explanation of life's origin, its primary issue was the same then as it was a hundred years later: How could organic substances come from the inorganic materials of the earth?

Berzelius's distinction between organic and inorganic chemistry stood only

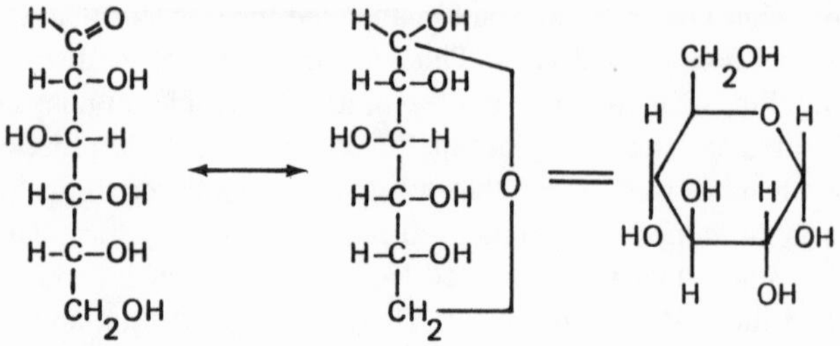

Figure 7.1. Structural formulae of D-glucose.

twenty-one years. One of his own students, the German chemist Friedrich Wöhler, brought down the barrier abruptly. It was a simple experiment—but indisputable. By heating the inorganic salt ammonium cyanate, Wöhler produced urea, the common substance excreted in mammalian urine. It was a hard blow to the Vitalists. They recovered somewhat to redefine their position, but Wöhler's experiment was only the beginning of the breakdown of the division between chemistry of the animate and of the inanimate.

Eleven years later, in 1839, Schwann swept all living things into a common mold by recognizing the cell as the basic unit of life. And for the remainder of the century, while biologists studied the composition and function of cells, chemists unveiled, step by step, the architecture of organic compounds.

In 1812, Gustav Kirchhoff astonished everyone by boiling ordinary starch with a little acid and obtaining grape sugar. He had hydrolyzed the starch, that is, ruptured the covalent bonds with the addition of water, by a kind of digestion catalyzed by the acid. Seven years later the French chemist Braconnet boiled several plant materials, including sawdust, and generated the simple sugar, glucose. It was subsequently learned that a substance named cellulose, a principal constituent of plants, was the source of Braconnet's glucose.

Joseph Louis Gay-Lussac investigated starch, cellulose, and the various sugars and discovered that each analyzed one carbon atom for each oxygen and two hydrogens. There appeared to be a water molecule for each carbon, giving the formula CH_2O for the members of this group. This prompted him to name them carbohydrates, or hydrated carbon.

The sugars were the simplest molecular structures obtainable by hydrolysis. When the molecular weight of glucose was measured, it was found to be 180, or six times the molecular weight of CH_2O. The molecular formula, therefore, must be its multiple by six, or $C_6H_{12}O_6$. Other simple sugars gave the same formula. Since starch and cellulose were considerably larger, a size immeasurable by techniques of that time, they were apparently formed by the linking of the simple sugars.

A Dutch agricultural chemist, Gerardus Mulder, studied the albuminous substance found in foodstuffs, and in 1858, upon a suggestion by Berzelius, applied the term *protein* to the group. Biologists quickly realized that protoplasm, the viscous solution found in all cells and regarded as the basis of life, was composed mostly of protein. Hydrolysis of the proteins, unlike that of carbohydrates, gave a mixture of smaller nitrogen-containing compounds that were difficult to separate. The first of this series to be isolated was found as a white crystalline chemical in 1820 and was called glycine. These are the alpha amino acids, distinguished by having a carboxylic acid and amino group joined to the same carbon atom, and named in 1848 by Berzelius. Over a century would pass before all of the twenty-one common amino acids of proteins were isolated and identified.

By 1827 William Prout was already suggesting that the organic matter of living systems was made up of essentially three classes of substances: the carbohydrates, lipids, and proteins. As chemists continued to investigate them, they learned that a characteristic of starch, cellulose, and proteins was that they were extremely large molecules. This distinguished them from inorganic compounds, which rarely exceeded a molecular weight of a few hundred. Measuring the size of these organic polymers was beyond the range of the methods available to the chemist of the mid–nineteenth century, but the scientists had discovered an important fact: life was constructed of giant molecules.

Berzelius, who had set chemistry on its course, died in 1848. That same year a young French chemist named Louis Pasteur presented to the Paris Academy of Sciences a paper telling of a remarkable discovery he had made. Certain chemical compounds existed as "right-handed" and "left-handed" components, each being mirror images of the other. Tartaric acid, discovered in particular industrial processes, had been found to have the same composition as the native acid from grape fermentation, but differed in a single property. This property was that natural tartaric acid viewed through a polaroscope gave a rotation to a beam of polarized light, whereas the synthetic acid failed to do this.

Pasteur studied the crystals of the synthetic tartaric acid under a microscope and discovered two crystalline forms, mirror images of each other. By painstakingly separating the crystals with forceps, he found that the two forms had opposite effects on polarized light. He established that this was due to molecular asymmetry and further showed that this property of optical isomerism—having the same molecular formula but different rotation of polarized light characteristics—occurred in other organic compounds, but that only one form existed in natural products. In chemical syntheses both forms were produced in equal proportions so that the substance exhibited no rotation of polarized light. Since only one isomeric form was found in nature, Pasteur deduced that organisms were selective toward isomers, a fact that prepared him for another discovery.

Until Pasteur's time yeast was not regarded as a microorganism, but merely an organic catalyst in the leavening of bread and the fermentation of fruit juices. When asked by the wine industry to investigate the problem of their product going bad, Pasteur studied fermentation and correctly guessed that the conversion of sugar to alcohol was the result of the biological activity of a living organism,

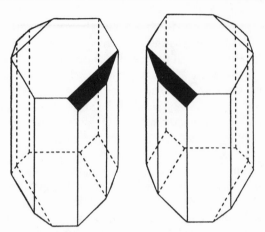

Figure 7.2. Drawings of crystals of tartaric acid showing
the mirror images of the two asymmetrical isomers.

the yeast. He further concluded that there was a specific microorganism that was
responsible for the fermentation of juices, the souring of milk, and the putre-
faction of meat. Through his monumental achievements in microbiology, Pasteur
brought to light a better understanding of life on the microscopic level. By a
simple experiment, he proved that each microbe was derived from a preexisting
microbe, thus giving even microorganisms an ancestry and finally putting to rest
the notion of spontaneous generation.

The chemistry of organic compounds entered its second period in 1858 with
the birth of the structural theory. By this time it was becoming increasingly more
difficult to reconcile so many molecular formulas for the growing number of
organic compounds. August Kekulé studied the problem and showed carbon to
be tetravalent, which is to say, carbon formed four chemical bonds with its
neighboring atoms. But more importantly, he recognized that carbon was able
to link not only with other elements but also with other carbon atoms, resulting
in long chains called the aliphatic series. Later, he postulated the aromatic series
consisting of molecules containing the benzene ring.

A new dimension was added in 1874 when Jacobus van't Hoff and Joseph
Le Bel proposed that the four valence bonds of carbon were directed toward the
four corners of a tetrahedron with carbon at the center. The architecture of organic
compounds became three-dimensional. Although the representation of carbon
bonds as a flat picture directed at right angles has been retained out of conve-
nience, it was realized that bonds oriented toward the corners of a tetrahedron
gave a spatial expansion to organic substances.

In 1875, Emil Fischer, who had been one of Kekulé's students, succeeded in
elucidating the three-dimensional structure of sugars and confirmed their structure
formulas by synthesis, generally the final confirmation of the correctness of a

formula. Since starch was glucose linked together in a polymer, its structure was then recognized to be

There were two basic types of starch molecule, the linear polymer and the branched. When the molecular weights were eventually determined, the linear polymer was found to vary in size from 100 to 1,000 glucose units.

The architecture of substances isolated from plants and animals was the largest and most complex that chemists had ever encountered. The powers of chemical synthesis exercised by biological systems were certainly impressive, even intimidating. Organisms produced natural products of immense size and extreme complexity like chlorophyll ($C_{55}H_{72}MgN_4O_5$) and hemoglobin ($C_{738}H_{1166}FeN_{202}S_2)_4$ with incredible ease. To synthesize even simple compounds in the laboratory, the chemist had to use elevated temperatures and highly reactive agents, and generally obtained only a low yield of the desired product. With the biological system, how was it possible?

The controversy raged for over fifty years during the last half of the nineteenth century between the Vitalists and the Mechanists. The Vitalists maintained that biological reactions were possible only because of the living organism; the Mechanists believed biological processes to be reactions of chemicals, albeit very complicated, but nevertheless, chemical compounds.

Back in the 1830s people had studied the digestive process and discovered hydrochloric acid in the extracts from the stomach. Since acid hydrolysis of carbohydrates and proteins was known, it seemed apparent that this was the process of digestion. Then in 1835, Theodor Schwann isolated from gastric juices an organic powder which was not an acid, yet was very active in breaking down meat. He gave it the name *pepsin* from the Greek word for digestion. Other organic catalysts were found. In 1833, Payen and Persoz isolated diastase from malt, which hydrolyzed starch to sugar, Justus von Liebig and Friedrich Wöhler reported in 1837 emulsin from bitter almonds, and Dufrunfaut extracted an organic catalyst from yeast which degraded sucrose to the simple sugars, glucose and fructose. Since each of the organic catalysts effected a hydrolysis, they were regarded by the Vitalists as agents of digestion, a process that can proceed outside the body. But to transform one organic compound into another the way yeast changes sugar to alcohol in fermentation, they argued, required the living organism.

In 1897, Eduard Büchner, a German chemist, attempted to analyze this process of alcoholic fermentation. He ground the yeast cells with sand until they were all fragmented, then removed the sand and cellular debris by filtering. When Büchner added sugar to the filtrate—now free of any yeast cells—rapid fermentation began at once. Fermentation did not need the cell. The biological reaction was being carried out not by the living organism but by some substance extracted from the cell. Büchner recognized that fermentation was not a physiological process but a chemical reaction catalyzed by an enzyme. He called the enzyme zymase.

Why did Büchner's experiment work where others failed? The answer lies in the fact that enzymes are frail structures. Always before, the method used to kill the yeast cells also destroyed the fragile configuration of the enzyme molecules.

Büchner's discovery established that the biological processes carried out by organisms were really chemical reactions catalyzed by organic components called enzymes that could be investigated separately without the cell. It was quickly realized that pepsin, emulsin, diastase, and invertase were also biocatalysts belonging to the enzyme class. Little was known, however, of the chemical makeup or mode of action of enzymes except that they were very large and unstable.

The encounter with molecular sizes too large to measure had always been an unsettling obstacle to the study of biological material. With no definite estimate of molecular size, the image of starches, proteins, and enzymes remained vague and mystical.

Advances in the measurement of large molecular weights were begun in 1877, when the German botanist Wilhelm Pfeffer attempted to use a principle discovered seventeen years earlier in Scotland by Thomas Graham. Graham had learned that a thin parchment between a salt solution of protein and pure water allowed the salt to pass through the pores but not the protein, the latter being too large. Actually, small ions of salt and molecules of water pass freely until they equilibrate with an equal number on both sides. But if one side contained a substance like protein which was too large to pass through the semipermeable membrane, then at equilibrium for water, the volume of the protein solution increased, since initially a unit volume of the protein solution had less water than a unit volume of water. This increase in volume was called osmotic pressure.

What Pfeffer realized was that osmotic pressure depended on the number of protein molecules, and since the number in the weighed sample depended on molecular size, osmotic pressure related to molecular weight. The procedure was refined later by the Dutch physical chemist van't Hoff. The difficulty was to chart a reliable correlation between the osmotic pressure one reads in the experiment and some known molecular weights. After the calibration of the method was accurately established, it was possible to measure the osmotic pressure of a weighed sample of any protein and compute its molecular weight.

The molecular weights of proteins were found to be huge. Egg albumin was

34,000, hemoglobin, 67,000, few proteins were less than 10,000, and many were larger than 100,000, the upper range to the method of measurement. This meant that egg albumin was a biopolymer of about 300 amino acids, and hemoglobin one with over twice that number. When osmotic pressure was superseded by the ultracentrifuge in the twentieth century as the means of determining molecular weights, the size of proteins larger than 100,000 came within the scope of measurement. But the possibility of ever knowing the order of sequence of all the amino acids and the complete shape of such giant structures appeared remote.

Enzymes were the tantalizing mystery of life. Neither their chemical nature nor mode of action was known. Since enzymes were such powerful catalysts, only minute amounts were needed to promote the specific chemical conversions. Consequently, it was easier to measure the result of enzymic activity than to detect the enzyme itself. Some enzymes, such as carbonic anhydrase, which dehydrates carbonic acid, can transform over a million molecules a minute for each molecule of enzyme. Analysis of an enzyme concentration required a measurable quantity of purified sample, a level of preparation that was difficult to achieve with the laboratory methods at the turn of the century. One thing that was certain was that enzymes were notoriously unstable and lost their activities quickly if warmed to only 56°C, just nineteen degrees above body temperature. It was an intriguing phenomenon and one reminiscent of proteins. Because of this, many biochemists began to believe that enzymes were really proteins with catalytic properties.

The controversy stormed on for and against this concept for a couple of decades, neither side being able to gather adequate data to settle the dispute. Then in 1920, Richard Willstätter, a renowned German chemist and Nobel Prize winner, studied the problem. He purified an enzyme sample with extreme care until he was satisfied that no other substance remained. The resulting clear solution still retained enzymic activity, despite the fact that no protein could be detected. Confident of his results, Willstätter announced to the scientific world that he was convinced that enzymes were not proteins. He was wrong. Willstätter failed to realize that enzymes could exhibit their activity at concentration levels far below the amounts needed to detect proteins by the methods known at the time.

Then in 1926 an incredibly simple experiment carried out by a biochemistry professor at Cornell University changed several ingrained impressions. James B. Sumner had been working for nine years trying to isolate the enzyme urease from jack beans. The crux of the problem was to find a means of selectively precipitating the urease from crude extract to isolate a purified product. One day, following a suggestion by his former professor at Harvard, Sumner used acetone to extract his jack bean meal. After allowing the solution to filter overnight, he examined a drop of the filtrate under a microscope. There were tiny octahedral crystals that he had never seen before. Collecting the crystals by centrifugation, he dissolved them and tested the solution. They showed intense urease activity. Subsequent work revealed the crystals to be a protein with a molecular weight of 483,000. Not only had Sumner shown that the enzyme was a protein, he had actually crystallized it. The substances that promote the fundamental processes of life were compounds that crystallize in the manner of common chemicals.[1]

8

The Molecular Basis of Life

The astonishing feature of living things is that they can reproduce facsimiles of themselves generation after generation. The conclusion is inescapable. Organi.ms must possess
within their makeup some means of retaining and passing on
a store of information that is the inheritance from the preceding generations. This information contains the instructions for synthesizing the organism and all its components.

Since prehistoric times man has realized that heredity was
an influence in the physical characteristics of plants and animals, but the mechanism remained obscure and mystifying.
Not until the middle of the last century was a systematic
study of inheritance carried out. In 1856, an Augustinian
monk named Gregor Mendel, growing varieties of the common pea in the cloister gardens in Brünn (Brno), Moravia
(now a part of Czechoslovakia), began experiments in cross-
breeding them and observing the transmission of various traits
in the offspring in the first, second, third, and following
generations. As early as 1866, Mendel published statistical
rules regarding inheritance in the Proceedings of the Natur-
forschender Verein in Brünn. The paper received little attention at the time and was forgotten, only to be rediscovered
in 1900 when three European botanists, Carl Erich Correns
(Berlin), Erich Tschermak von Seysenegg (Vienna), and
Hugo De Vries (Leiden), simultaneously and independently
reported results similar to Mendel's, only to find that the
experimental data and theory had been published 34 years
previously.

The great controversies over evolution raged throughout
the latter half of the nineteenth century and into the twentieth. In order to explain Darwin's theory, biologists for-
mulated the concept that biological characteristics are

inherited by physical factors that are passed on through successive generations. The English biologist William Bateson gave this branch of biology the name *genetics* (from genesis) in 1906, and in a peculiar retrogressive derivation, the inheritance factors came to be known as genes. But what is the actual chemical nature of the gene?

The genetic substance was isolated from the nuclei of cells nearly 70 years before its true biological significance was realized. In 1868, Friedrich Miescher, a young medically trained Swiss chemist from Basel on the borders of France and Germany, set out for Tübingen. At twenty-four he had just completed his doctoral examination and was going to work for Ernst Hoppe-Seyler, the great German physiological chemist. It was autumn before Miescher began his postdoctoral research, but by February of 1869 he wrote to his former professor in Switzerland of isolating a new substance from cell nuclei.

Little was known of the nucleus of the cell at this time, and the function of cellular material was almost completely obscure. Miescher had initially planned on carrying out his investigation on lymph cells, but their limited availability compelled him to use pus cells that he extracted from surgical bandages. Contaminated with grease and carbolic acid, the cells were first washed with sodium sulfate solution, filtered, and treated with alkali. Miescher then shook the cell fragments vigorously for a long time in mixture of ether and extremely dilute hydrochloric acid. The fats, decomposition products, and detritus either dissolved in the ether or went to the interface of the immiscible liquids: the slightly denser nuclei slowly settled and fell to the bottom of the water layer as a fine, whitish sediment.[1]

Miescher's substance from the nuclei of pus, which he called nuclein, contained a substantial percentage of phosphorus. Until this time, lecithin was the only phosphorus-containing natural product known. The nuclein was a complex of protein and nucleic acid; but subsequent purification procedures led to the nucleic acid being separated as a mass of long, fibrous, threadlike material that could be collected from the precipitated matter by entwining it on the end of a glass rod. Apparently Miescher, without knowing the structural nature of his nuclein, realized it had some connection with the genetic function. It was not until 1944, however, that an experiment by O. T. Avery[2] verified that the young Swiss had isolated the substance that is the chemical basis for the hereditary features of all living things—the nucleic acid.

Miescher's preparation of nuclein came ten years after Darwin published *On the Origin of Species*. By 1880 the mitotic process was established, and biologists, working with the light microscope, discovered that all cells contain nuclear material in a definite number of rodlike units called chromosomes. Chromosomes are the carriers of specific hereditary factors (genes), and elegant microscopic studies revealed their mode of replication in cellular reproduction. When isolated and analyzed, chromosomes were found to contain protein and nucleic acid in nearly equal proportions.

Nucleic acid is the phosphorus-containing polymeric substance found in the nuclein that Miescher had isolated from pus cells in 1868. The German chemist Albrecht Kossel, working at Heidelberg with nucleic acid from the thymus glands of calves, discovered that it contained nitrogenous bases called purines and pyrimidines, and he isolated and identified two different derivatives of each base. Being a polymer with only four kinds of subunits, nucleic acid, like starch, seemed to be a long, monotonous chain molecule.

For the first few decades of this century, while biologists expanded genetics by clarifying the mathematical relation of biological inheritance, biochemists achieved considerable success in showing the role of enzymes in controlling the life processes. Neither had any clear concept of the chemical nature of a gene or how enzymes were made. Then in the 1930s, the American geneticists George Beadle and Edward Tatum,[3] working with *Neurospora crassa,* the common red bread mold, linked Darwinism to chemistry by showing that enzymes control structure and genes control enzymes. They recognized that genes are somehow coded for enzymes, and they postulated that for each enzyme there is a specific gene.

It was an impressive feat for cells to reproduce proteins containing hundreds of amino acids repeatedly in the exact sequence. For a protein to be synthesized time and again in identical composition, it was reasoned, the amino acids must be polymerized in their precise order on some type of template. Such templates would contain in their composition the information of heredity.

Because of the immense amount of information an organism would need in its heredity material, biologists generally accepted that for any substance to be the genetic factor it would have to have a large number of subunits to act as letters in the informational code. Since proteins contain over twenty kinds of amino acids, these macromolecules appeared ideally suitable for the role; and throughout the 1930s, most biologists and biochemists believed the protein found in chromosomes was the genetic material and regarded the nucleic acid to be of little significance.

During that same decade, in 1935, Wendell Stanley[4] at the Rockefeller Institute in New York did an astonishing thing. He crystallized a virus. Viruses are biological entities that occupy a zone between living cells and inanimate chemicals. They consist of protein and nucleic acid without the cellular machinery for reproduction and metabolism. Lacking the means of regeneration, viruses are perpetuated by inserting their genetic factor into cells, and the machinery of the infected cell is taken over to produce copies of the virus. For a virus to be crystallized like so much salt impressed upon scientists that genes might eventually be isolated and studied as chemical compounds. The basis of life moved one step closer to being purely a matter of chemistry.

Then in 1944, Erwin Schrödinger,[5] the renowned Austrian-born physicist living in Dublin as an émigré from Nazi Germany, published a small book, *What Is Life?* In it he stressed upon his fellow physicists that biology was on the threshold of the crucial question of the basis of life, and they must not be discouraged by

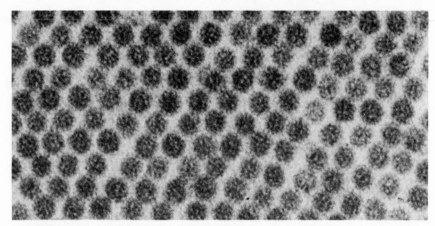

Figure 8.1. Electron photomicrographs of human wart virus. The virus is an icosahedral shell containing DNA. This particular virus has no envelope. Magnification is 120,000 X.

the difficulty in interpreting life by the ordinary laws of physics. He emphasized that they should consider finding how biology can be explained on the molecular level.

The key to the puzzle lay in viruses that infect bacteria and are called bacteriophages. In 1952, Alfred Hershey and Martha Chase, using radioactive phosphorus and sulfur as labels to follow the respective biochemicals, demonstrated that the DNA of a bacteriophage entered the bacterium and it alone was responsible for the reproduction of new viruses. This compelling evidence that the chemical form of genes was the nucleic acid astounded biologists, who regarded protein to be the material of inheritance. The discovery marked the beginning of molecular biology.

When Kossell analyzed nucleic acid in the last century, he found two purines he called adenine and guanine, and two pyrimidines, cytosine and thymine. Later research with plant nucleic acid led to the discovery of a third type of pyrimidine called uracil (fig. 8.2).

In 1910, Levine at the Rockefeller Institute found that nucleic acids also contained a five-carbon sugar. The nucleic acid from plants had the sugar ribose, whereas animal nucleic acid had the same sugar minus one oxygen, and hence was known as deoxyribose (fig. 8.3).

The structural units of nucleic acid consist of the purine and pyrimidine base bonded to the terminal carbon atom (no. 1) of the sugar, and the sugar portion has a phosphate group attached. These three constituents—base, sugar, phosphate—together form a nucleotide. When nucleotides join through a phosphate diester by a 3', 5'-linkage of their sugars, they create the long chains known as nucleic acids (fig. 8.4).

It soon became apparent that there were two kinds of nucleic acids: one contained adenine, guanine, cytosine, and thymine with deoxyribose and was

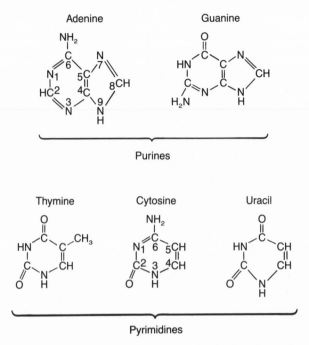

Figure 8.2. Structures of adenine, guanine, thymine, cytosine, and uracil.

called deoxyribonucleic acid (DNA) (fig. 8.5); the other nucleic acid also consisted of four bases but contained uracil in place of thymine, and, since the sugar of uracil was ribose, it was called ribonucleic acid (RNA).

Eventually, biologists realized that the two kinds of nucleic acid did not distinguish plants and animals, but that all living things contained both types, DNA and RNA. Beneath life's immense diversity there was an astonishing unity. The principles of genetics were found to be coextensive to all forms of life. The

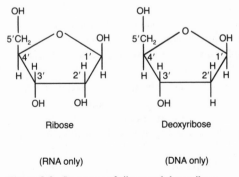

Figure 8.3. Structures of ribose and deoxyribose.

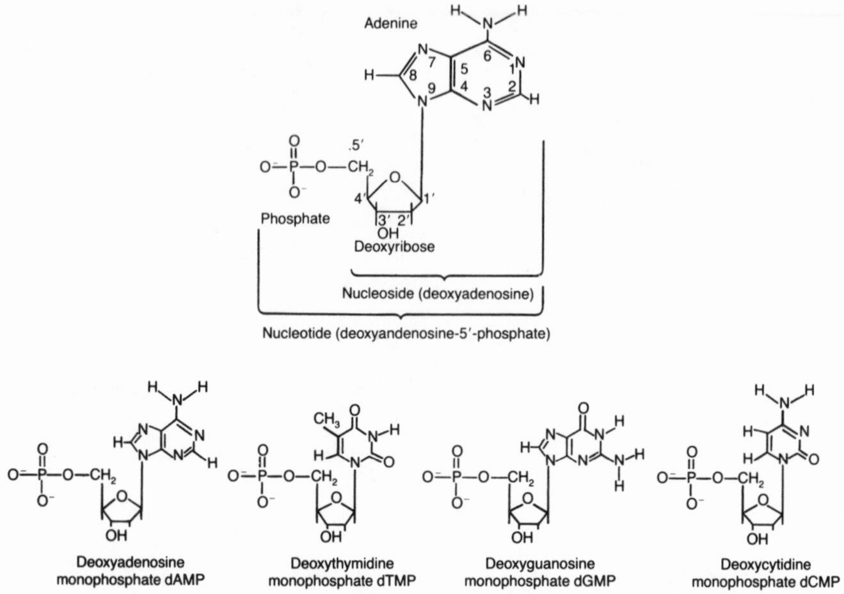

Figure 8.4. Structures of deoxynucleotides.

continuity of living substances through reproduction is based on the multiplicity of genes which are copied and passed on from generation to generation—the material of inheritance being the same for viruses and man alike. That substance is DNA.

Only four basic chemical units are used to create the blueprint in a coded sequence of units in an informational molecule where the molecular weight can run into billions even for bacteria. A section of the DNA constitutes a gene that carries the information for the amino acid sequence for a particular protein. In bacteria the number of genes can be in the thousands, but in mammals it runs as high as 100,000.

Cellular reproduction is ultimately molecular reproduction, and the unique nature of the chemical structure of nucleic acids allows these biopolymers to be copied faithfully. In procaryotes, with their circular DNA, cellular reproduction is replication of the nucleic acid, followed by binary fission into daughter cells with each carrying a full complement of cellular material. The chemistry is simple and direct and must represent the mode of replication assumed soon after the origin of the first living systems. Eucaryotic cells, on the other hand, have reproduction procedures so elaborate that they must have taken an extremely long time to evolve.

For DNA to be the informational reservoir of the cell, there had to be a chemical procedure for making duplicates of the molecule. Before understanding

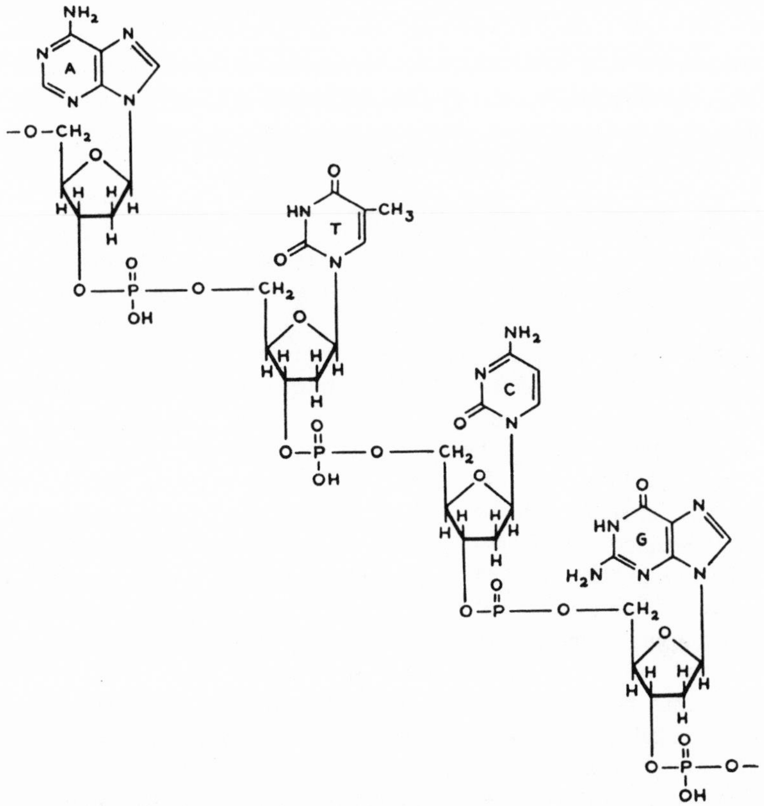

Figure 8.5. A tetranucleotide portion of one strand of DNA composed of adenine (A), thymine (T), cytosine (C), and guanine deoxynucleotides.

how this was accomplished, it was necessary to determine the complete structural arrangement. Nucleic acids are polynucleotides that exist as long, unbranched chains, but only the precise three-dimensional configuration would reveal the mode of action of their biological function. And thus was launched the search for the secret to DNA's architecture.

In 1950, Erwin Chargaff and his students[6] at Columbia University made an odd discovery. After carefully analyzing the purine and pyrimidine composition of various DNAs, they noticed that the number of adenine bases nearly always equaled the number of thymines, and the number of guanine and cytosine bases were also nearly equal. To put it more conveniently, they found A = T and G = C. This was despite large variations in the amounts of A = T to G = C in different DNAs.

While Chargaff's laboratory was analyzing the base composition of DNA, Rosalind Franklin and Maurice Wilkins at King's College, London, were using X-ray crystallography to obtain some precise measurements of DNA. And in

California, Linus Pauling and his associates[7] at the California Institute of Technology attacked the problem of DNA's structure by using the bond lengths and angles of the quantum theorists as guides to construct atomic models. Both the London group and the Caltech chemists were inclined to believe that there were three polynucleotide chains in the DNA molecule.

The principal force of attraction that would hold the polynucleotide strands together was hydrogen bonding. Hydrogen bonds are weak compared to covalent bonds. To break a covalent bond requires between 12 and 24 KJ/mole,* whereas the energy to break hydrogen bonds ranges from 1 to 3 KJ/mole. Nonetheless, hydrogen bonds are of enormous importance in biology by being primarily responsible for the specificity of interactions between macromolecules.

A critical question concerning the structure of DNA was whether the bases pointed toward the outside or toward the center of the molecule. Pauling suggested that the bases were on the outside, but Franklin felt that she had evidence that the phosphates were on the outside and the bases were toward the center.

This was the situation in 1951 when James Watson, a twenty-two-year-old American postdoctoral fellow, arrived in Cambridge and met Francis Crick, a physicist working on his Ph.D. in biophysics. Although ostensibly Crick was doing research on protein and Watson was interested in the structure of the Tobacco Mosaic Virus, they both had an abiding interest in DNA and decided to collaborate on solving the riddle of its structure.

Franklin's X-ray diffraction patterns indicated a regular and compact configuration. The helical structure had been shown by Pauling to be a favored configuration of macromolecules, and Crick became enamored with helices. The question of prime importance was the chemical basis of the procedure organisms used to copy their DNA. Presumably the purines and pyrimidines were important. Reasoning that nature has a penchant for doing things in pairs, Watson discarded the three-strand concept and felt it more plausible to build a model with two strands twisting around each other with the bases directed toward the center. Using cardboard cutouts of the four bases, he attempted to build a model on hydrogen bonding with each base facing a like base as it could be imagined duplication of a strand to occur. But the model was not compact as shown by the X-ray studies. And as Jerry Donahue, an American crystallographer working in the same laboratory, pointed out, Watson, like everyone else, was using the wrong tautomeric form of the bases. Donahue felt that the bases existed in the keto rather than the enol form.

With new cutouts, Watson again attempted to combine base pairs for his model. At this point he discovered that the cutouts of adenine with thymine were the same size and shape as guanine with cytosine. The significance of Chargaff's A = T and G = C ratios suddenly became obvious.

---

* Kilojoules. One kilojoule = 0.239 kilocalorie.

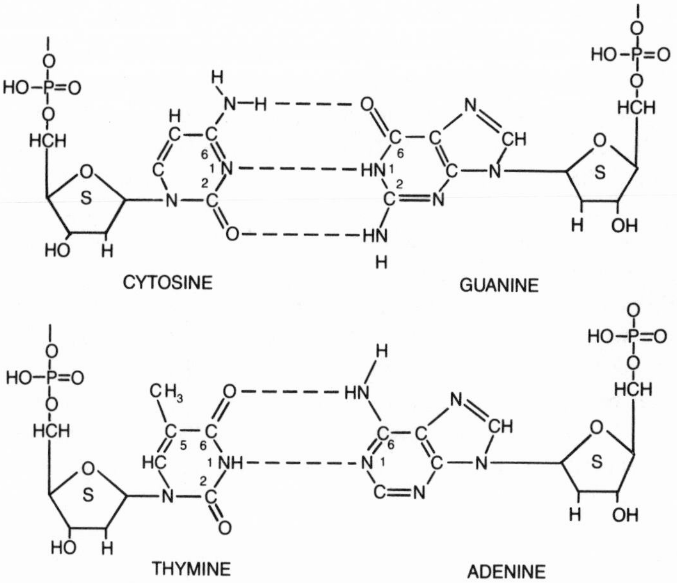

Figure 8.6. The arrangement of nucleotide pairing in DNA.

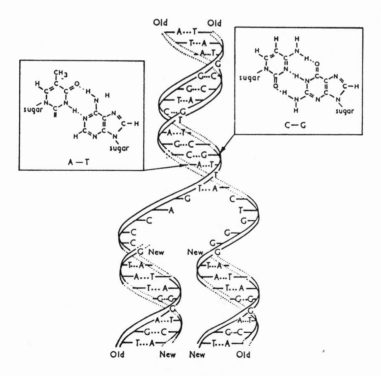

Figure 8.7. Illustration of unwinding and replication of the DNA molecule.

Since it was possible for the adenine–thymine and guanine–cytosine pairs to be held together by at least two hydrogen bonds to give nearly identical shapes, the model would be consistent with the compact arrangement indicated by the X-ray analysis. Instead of replicating through a like–like combination, the model showed the DNA molecule to be two strands of polynucleotide chains twisted in a double helix held together by bonding between complementary bases, with one strand acting as the "negative" and the other as the "positive." Replication was achieved upon unwinding when each single strand became the template for the new complementary strand (fig. 8.7).

Within eighteen months, Watson and Crick had constructed a model of DNA consistent with the physical data that showed not only the three-dimensional shape of the DNA, but also the manner in which the molecule is duplicated. Since their results were first published in 1953,[8] a great deal of evidence has accumulated supporting their proposed structure, the most spectacular being Arthur Kornberg's[9] synthesis of a biologically active DNA molecule in the test tube.

9
From Blueprint to Organism

Miller's experimental simulation of atmospheric conditions on primordial earth opened the research to how life began by showing the manner in which the building blocks were formed before the first cells. Among the building blocks are the nucleotides that join to form nucleic acids, and it is the nucleic acid DNA that is the molecular basis of life's reproductive capacity. But DNA is only an informational molecule—like a computer tape. The proteins, in their immense variety and roles, are the chemicals most directly responsible for the shape, the composition, and the functionality of an organism. Only after the information coded in the chemical structure of DNA is retrieved and translated to protein structure does it complete its biological purpose. How then are proteins created from the DNA structure?

There are two kinds of nucleic acids: DNA and RNA. DNA is the molecule that is the store of hereditary information and is found in the nucleus of the cell. RNA, on the other hand, is found both in the nucleus and in the cytoplasm. Protein synthesis takes place in the cytoplasm. Even in the 1940s, before DNA was known to be the material of genes, research by Torbjörn Caspersson[1] in Stockholm and Jean Bracht[2] in Brussels indicated that RNA somehow was involved in protein synthesis.

In 1950, Henry Borzook and his colleagues[3] at the California Institute of Technology and Tore Hultin of the Wenner-Gren Institute in Stockholm independently identified the microsomes, later known as ribosomes, in the cytoplasm as the site of protein synthesis. The DNA does not act directly as the template for the synthesis of protein, but instead, the DNA sequence is transcribed to an RNA molecule which is copied from the DNA in the same manner that

DNA is replicated. The RNA copy of DNA is a messenger RNA (mRNA) that carries the hereditary information in its sequence to the ribosome to be translated into the amino acid sequence of a polypeptide.

$$\text{DNA} \xrightarrow{\text{(transcription)}} \text{mRNA} \xrightarrow[\text{ribosome}]{\text{(translation)}} \text{protein}$$

In the mid-1950s biologists thought ribosomal RNA was literally a template for protein synthesis which contained structural cavities that were complementary to the shape of amino acids from which the protein was constructed. But no one could construct a model of RNA with specified holes that could conceivably act as a template. It was Francis Crick who recognized that amino acids do not fit directly in the template molecule but require an adapter molecule that is specific for the amino acid that recognizes the particular site on the template.

Mahlon Hoagland of Harvard University had discovered a type of RNA that could not be settled by centrifugation but remained in solution. Hence he called them soluble RNAs. It was not long before molecular biologists realized that Hoagland's soluble RNAs were Crick's adapter molecules. Within a short time a specific soluble RNA was found for each of the twenty amino acids, and more in keeping with their role, they became known as transfer RNAs (tRNAs).

The sequencing of tRNA molecules showed that all of them had CCA at the end of the polynucleotide that has the sugar with a free dydroxyl group on the 3'–carbon—the end to which the amino acid is attached. Also, since the A/U and G/C ratio approached equality, there was the possibility of considerable base pairing within the molecule. Robert Holley and his coworkers[4] at Cornell University observed that tRNAs contain a number of unusual bases that differ from adenine, guanine, cytosine, and uracil by having one or more methyl groups substituted at various positions in their structures.

A complete analysis of a tRNA for alanine was accomplished by Robert Holley and others in 1945, and the nucleotide sequence of a yeast tRNA for tyrosine was reported by Madison and coworkers[5] the following year. The structures proposed for these two tRNAs was a cloverleaf configuration held together by hydrogen bonding and with the CCA bases at the 3' position and sticking out. Since then, sequences of some 75 different tRNA molecules have been determined, and they all can be organized in the same general cloverleaf folding.

The middle loop of the cloverleaf on the opposite end from the CCA contains a triplet of unpaired bases that form the anticodon, these three bases being complementary to the triplet of bases in the mRNA that represent a codon, or one word in the information in the mRNA that translates to one amino acid. The tRNAs range in number of nucleotides from 74 and 91, but of the various tRNA molecules, the overall distance from the CCA at one end to the anticodon at the other end appears to be constant, the difference in the nucleotide number being compensated for by the size of the little loop located between the right hand and

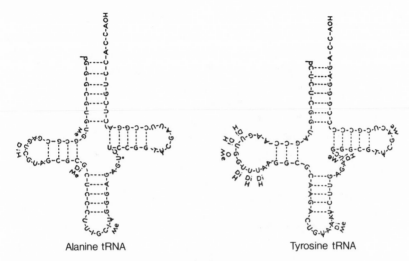

Figure 9.1. Proposed structure for alanine tRNA and tyrosine tRNA.

bottom limbs. Also, the unusual bases are located in regions not forming the hydrogen bonds. Transfer RNAs, unlike the other forms of nucleic acids, must have a highly specific three-dimensional structure, and in this respect they actually resemble proteins.

Amino acids are prepared for use in protein synthesis by being activated in a reaction with adenosine triphosphate (ATP), which requires the enzyme aminoacetyl RNA synthetase. The result is an activated complex of enzyme-AMP-amino acid in which the 5′-phosphate group of adenosine monophosphate (AMP) is linked as a mixed anhydride to the carboxyl group of the amino acid. Pyrophosphate is split from the activating ATP in the process.

$$
\begin{array}{c}
\text{R--CH--COOH} \\
\quad | \\
\quad \text{NH}_2
\end{array}
\xrightarrow[\text{Aminoacyl-RNA-}]{\text{ATP} \qquad \text{PP}_i \text{ (Pyrophosphate)}}
\text{E--adenine--ribose--O--}\overset{\text{O}}{\underset{\text{OH}}{\overset{\|}{\text{P}}}}\text{--O--}\overset{\text{O}}{\overset{\|}{\text{C}}}\text{--CH--R}\underset{\text{NH}_2}{}
$$

Amino acid   Synthetase (E)   Activated amino acid

In the next step, the activated amino acid molecules are transferred to the corresponding tRNA by another enzyme. There is a particular tRNA for each amino acid, and a specific enzyme for each amino acid to be attached to the tRNA. The tRNA complex containing the amino acid then moves to a ribosome.

The mRNA, having been formed from the DNA and moved to the cytoplasm, has ribosomes attached to it at the chain initiator codon (AUG), which binds, for example, N-formylmethionyl-tRNA. The ribosome moves down the chain of the mRNA one codon at a time, accepting the respective tRNA complex that corresponds to the codon and connecting the amino acid to the growing peptide chain. As the ribosome moves along the mRNA, the polypetide chain grows. Other ribosomes attach themselves to the mRNA as the initiator codon is vacated. In this way a single mRNA can serve as many as six or eight ribosomes at the same time. There is only one specific site of the ribosome where peptide bond formation can occur, but from each ribosome moving along the mRNA, a polypeptide chain grows, each on an average between 125 and 400 amino acids long.[6]

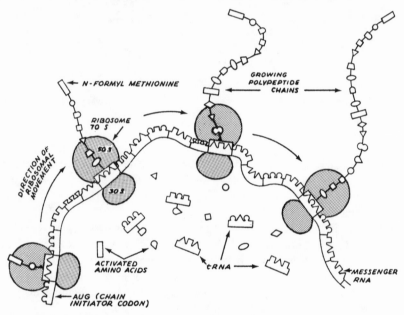

Figure 9.2. Illustration of ribosomes on a mRNA synthesizing polypeptides.

The sequence of purine and pyrimidine bases on the mRNA strand, originally transcribed from the nuclear DNA, directs the arrangement of amino acids in the synthesis of proteins. This information carried on mRNA is not in single nucleotides but resides in sequences of three nucleotides, or triplets. The "genetic code," therefore, consists of nonoverlapping triplets in the mRNA chain, each triplet being a code word, or codon.

After the resolution of the question of how hereditary information was stored in DNA, then transcribed to mRNA, and translated to protein structure, it remained to decipher the code. What nucleotides correspond to what amino acids?

M. W. Nirenberg and J. H. Matthael[7] in 1961 were the first to decipher a codon. They discovered that a synthetic polynucleotide containing only uridylic acid (polyuridylic acid) could serve as a template for ribosomes isolated from the bacterium *E. coli*. When poly U was mixed with ribosome preparations and tRNA molecules were added, each bearing its own specific amino acid, a polypeptide was synthesized that contained only phenylalanine. From this they were able to conclude that the codon for phenylalanine is UUU.

In similar experiments, GUU directed the synthesis for valine-containing polypeptides; UGU was the code word for cysteine, and UUG for leucine. Eventually, by the use of "block copolymers"—polynucleotides of repeating sequences of two or three bases—the entire genetic code was deciphered.

SECOND LETTER

FIRST LETTER	U	C	A	G	THIRD LETTER
U	UUU } Phe UUC } UUA } Leu UUG }	UCU ⎫ UCC ⎬ Ser UCA ⎪ UCG ⎭	UAU } Tyr UAC } UAA OCHRE UAG AMBER	UGU } Cys UGC } UGA UMBER UGG Tryp	U C A G
C	CUU ⎫ CUC ⎬ Leu CUA ⎪ CUG ⎭	CCU ⎫ CCC ⎬ Pro CCA ⎪ CCG ⎭	CAU } His CAC } CAA } GluN CAG }	CGU ⎫ CGC ⎬ Arg CGA ⎪ CGG ⎭	U C A G
A	AUU ⎫ AUC ⎬ Ileu AUA ⎭ AUG Met	ACU ⎫ ACC ⎬ Thr ACA ⎪ ACG ⎭	AAU } AspN AAC } AAA } Lys AAG }	AGU } Ser AGC } AGA } Arg AGG }	U C A G
G	GUU ⎫ GUC ⎬ Val GUA ⎪ GUG ⎭	GCU ⎫ GCC ⎬ Ala GCA ⎪ GCG ⎭	GAU } Asp GAC } GAA } Glu GAG }	GGU ⎫ GGC ⎬ Gly GGA ⎪ GGG ⎭	U C A G

[a] UAA (ochre), UAG (amber), and UGA (umber) are chain terminating codons. AUG is used as a chain initiating codon standing for formylmethionine in E. coli. In the middle of a protein chain it stands for methionine.

Figure 9.3. The genetic code.

As a result of many additional studies, not only were the triplets for all the amino acids assigned, but codons for special instructions were also found. When biologists discovered that all proteins synthesized by *E. coli* have N-formylmethionine as the N-terminal amino acid residue, they recognized that AUG was the initiator codon. All polypeptides begin with N-formylmethionine, which may or may not be removed later. Three amino acids have six codons, five have four codons, and ten have two codons. UAA, UAG, and UGA are chain-terminating triplets that signal for the peptide to end at that point.

Unlike proteins, nucleic acids are robust molecules. They are stable in mild acids and alkalies and can be heated to almost 100°C, conditions that wreck the delicate protein structures. Since the other biopolymers can be hydrolyzed without disrupting the nucleic acids, this property facilitates the isolation of DNA and the RNAs from the various cellular components. Another feature that simplifies the purification of DNA is that, in contrast to protein, there is only one kind of DNA in an organism.

Both DNA and proteins are biopolymers, built of subunits that evolved for different purposes. Proteins require shape and different types of functional groups to achieve specificity and chemical activities. To obtain these they needed a variety of subunits, and they found it in the twenty or so amino acids. DNA, on the other hand, did not need many shapes—it needed only to store information. For this purpose it had a choice. It could use a large alphabet (subunits) and have a large vocabulary, or it could use just a few letters but make the informational molecule extremely long. In the end, biological systems adopted a DNA of just four subunits and used three-letter words with a vocabulary of only 64 words. But the litany of life's message seems endless, for the number of units in DNA molecules often runs into the billions.

It can be argued that this is all that DNA needed to accommodate twenty amino acids. But there are many more possible amino acids than twenty that could have been adopted for proteins. Certainly, a DNA constructed of four kinds of nucleotides seems adequate and could have been the type that proved superior to other more complex arrangements by being simpler, more stable, and compact. And there may be no appreciable advantage in using more kinds of amino acids if twenty is enough.

The answer, however, may be for another reason. The emerging cell would have constructed its proteins and nucleic acids from the selection of amino acids and nucleotides in the primordial environment. Once committed to an efficient biological system, however, the primitive cells would have been unable to incorporate any additional building blocks. The amino acids and nucleotides adopted may have been the only ones available in the prebiotic environment of primordial earth.

10
A Thread Unbroken

All living things on earth are tied to an invisible evolutionary thread that stretches back to the beginning of life. The thread is the DNA molecule that is in each of our cells and carries the genetic information for the construction of our very being. That molecule has existed, has been altered, lengthened, and copied generation after generation from the moment the first living cells formed on earth over 3.4 billion years ago. The infrequent changes that occurred in the molecule were retained and passed down to succeeding generations; and each change in DNA became reflected in a change in a protein translated from its structure.

There are many kinds of proteins in a functional cell performing or monitoring essentially all biochemical reactions; a single cell can have 5,000 to 10,000, and man may have as many as 100,000. The metabolic processes, the matrix for bone and shell, the transport of reactants, the synthesis of constituents, and even the shape and mechanical properties of biological systems are all governed by proteins, each protein delicately adapted to perform its specific role by its chemical structure, which in turn is a direct expression of the arrangement of different amino acids in its chain. When, therefore, a change in the amino acid composition of a protein occurs, it often has a profound effect on the entire cell or organism. Nevertheless, this is the basis of evolution, and without it, life could not have developed.

It is the mutability of the genetic apparatus that results in alteration of proteins. A gene is a segment of the DNA molecule and carries the coded message for the synthesis of a single polypeptide—a polypeptide that may itself be a protein or may be combined with other polypeptides to form a protein. If, then, by mutation a nucleotide is changed in the

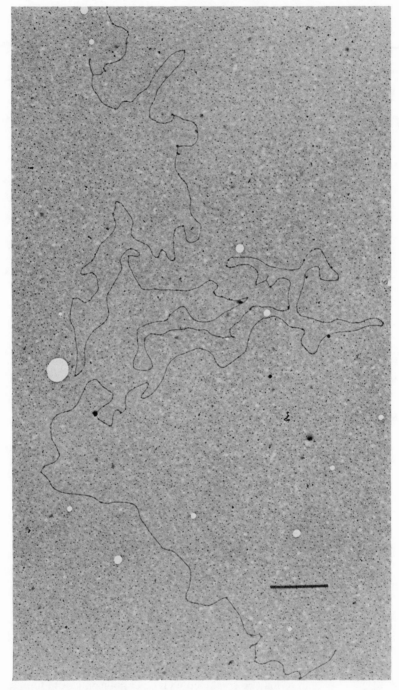

Figure 10.1. A molecule of linear double-stranded DNA. The bar represents 1,000 nm; magnification is 20,000 X.

gene, one of the coded messages is altered and is read for a different amino acid, so that when the protein is synthesized a substitution occurs in the peptide chain. Mutations generally result from the low background level of natural radioactivity, but they can also be caused by certain chemicals.

This is mutation on the simplest level and for single polypeptides. When advanced plants and animals evolved, their DNA molecules became clustered in chromosomes and reproduction became more complex. This allowed for greater variation of the genetic material. The diversity of genotypes can be attributed to different allele combinations, chromosomal interchange, inversion, recombination, or polyploidy. As a result, except in the case of twins, no two individuals of advanced species are identical. This complexity of reproduction introduced a new set of mutations by errors that occur in the procedure, but the mutation of the DNA molecule resulting in an amino acid substitution in a single protein is the subject of interest to molecular evolutionists. Chemically, it is simpler and can be studied analytically and mathematically.

The rate of mutation, like that of radioactive disintegration, is a statistical factor that is impressively constant. E. Zuckerkandl and L. Pauling[1] estimated that each amino acid in hemoglobin undergoes substitution by genetic mutation at an average rate of once every 800 million years. Since hemoglobin has 140 amino acids, this averages to one substitution for the molecule every 5.07 million years.

The constancy of the rate of substitution was confirmed by Motoo Kimura[2] by comparing the number of amino acid substitutions that have occurred in hemoglobin chains of man and the carp. Hemoglobin consists of two sequences of polypeptides called alpha and beta chains that evolved from an ancient hemoglobin of only one chain. Man and carp, with their two-chain hemoglobins, had a common ancestor with the primitive globin that lived during the Devonian period 350 to 400 million years ago, and divergence of the separate evolutionary lines is taken to have occurred around 375 million years ago. By comparing the alpha and beta chains of human hemoglobin which come from separate genes, Kimura found that they differed in the total number of amino acid substitutions by 75. When he then compared the number of substitutions of the human beta chain and the carp alpha chain, he found the difference to be 77, or essentially the same. In other words, after diverging from a common ancestor 375 million years ago, man and carp underwent virtually the same number of mutations in the alpha chain of their hemoglobins.

Various species have many of the same proteins in common which perform the same function, but differ slightly from species to species by amino acid composition. Such proteins are called homologs, and the more closely related species are, the more similar are the homologous proteins in composition. Vernon Ingram[3] of the Massachusetts Institute of Technology suggested in 1961 that the substitution rate in these homologous proteins by mutation could be used as a molecular evolutionary clock.

The degree of divergence in the composition of blood proteins in different animals is particularly evident in the study of blood serum. When human blood is injected into a rabbit, the rabbit's immune response produces antibodies to the human proteins. This antihuman serum in return, when mixed with human blood, causes clumping of 100 percent of the blood protein. The same antihuman serum mixed with blood of other species gives the following percent precipitation: gorilla, 64; orangutan, 42; baboon, 29; ox, 10; deer, 7; horse, 2; and kangaroo, 0. This is an extremely critical measure of chemical resemblance and affinity, and the order of relationship correlates with evidence from comparative anatomy, embryology, and paleontology.

Molecular evolution can now compare this divergence of amino acid composition of homologous proteins directly. Since 1965, the known sequences of proteins have been published annually by Margaret Dayhoff and her colleagues of the National Biomedical Research Foundation in the *Atlas of Protein Sequence and Structure*. The number of entries has grown from 50 protein sequences in the 1965 edition to 409 in 1972.[4] By the known sequences of homologous proteins from a number of species, a comparison can be made to assess to what extent each species has undergone amino acid substitution in the protein by mutation.

One of the earliest proteins to be studied was cytochrome c by Emanuel Margoliash of Abbott Laboratories and Walter M. Fitch of the University of Wisconsin Medical School.[5] This heme protein is easily extracted and purified and is found in every mitochondrion of eucaryotic cells, which is to say, of all living things above bacteria and blue-green algae. It therefore is the expression of a distinct, recognizable gene that has existed for some 1.3 billion years.

Cytochrome c in the vertebrates has 104 amino acids, a few more in others. Because the structural shape of cytochrome c is closely associated with its ability to perform by interacting with cytochrome oxidase and cytochrome reductase, the surface of the protein is conserved. Any mutation whose amino acid substitution changes the protein's shape is deleterious, and consequently, does not survive. As a result, 50 percent of the molecule remains invariant. Within the variable amino acids, there are 19 particular ones that play a role in the conformation of the polypeptides, and 6 others that have remained the same for undetermined reasons. There is also a constancy of the hydrophobic and basic segments of the protein, and some classes of amino acids are mutably interchangeable.

Thus, in studying the mutation distance between cytochrome c's of 20 species ranging from yeast to man, Fitch and Margoliash found a connection between the taxonomic kinship of species and the number of residue differences. Proteins of closely related species showed few or no differences, whereas the greatest differences in amino acids was with the most diverse species: the monkey and human lines are distinguished by a single mutation in the human line that resulted in the substitution of isoleucine for threonine at one position; and the horse differs from the tuna fish by 20 amino acids, from Baker's yeast by 45. The average difference between primates and the other mammals is 10.1 residues. With the amino acid substitutions from mutation through the history of cyto-

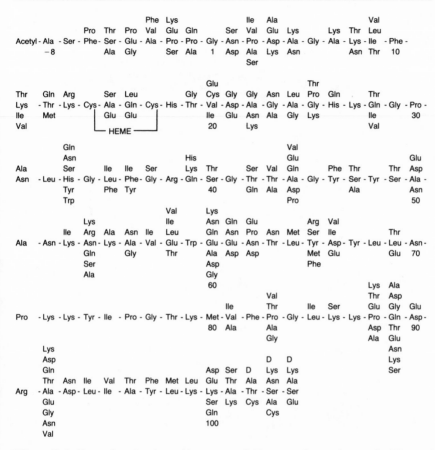

Figure 10.2. Composite of amino acid sequence of 30 eucaryotic cytochrome c's. The continuous sequence of 112 residues is that of wheat germ protein. All the other cytochrome c's are shorter with their various amino acids shown above and below the linear sequence.

chrome c calculated, Fitch and Margoliash were able to reconstruct the phylogenetic tree. Although substitutions in cytochrome c may not have been the mutation most responsible for a particular divergence, its evolution coincides remarkably with the taxonomic development of the species.

To derive the topology of a phylogenetic tree, the amino acid sequences of homologous proteins are programmed in a computer. The computation is based on the premise that the fewest number of changes occurred in the ancestral organism. The computer then considers each amino acid along the polypeptide chain, one at a time. Amino acids that have remained unchanged have been conserved and are considered invariant. For the others, the method involves a mathematical formula for each possible combination to take into account possible sequences for the ancestral structures. The detailed procedure is described in chapter 2 of the *Atlas of Protein Sequence and Structure* (1969).

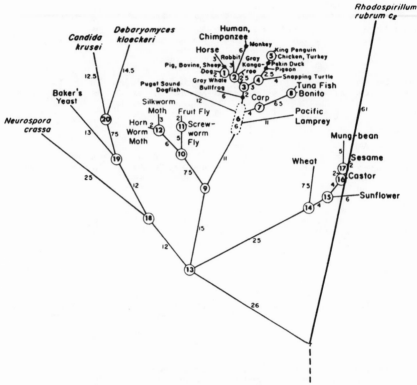

Figure 10.3. Phylogenetic tree based on cytochrome c. The numbers are the inferred amino acid changes per 100 links. (*Atlas of Protein Sequence and Structure*, 1972).

Motoo Kimura and Tornoko Ohta[6] have tabulated some of the principles governing molecular evolution. The rate of amino acid substitution is approximately constant if the change does not affect the function or tertiary structure of the protein. Therefore, those parts of the protein that are functionally least important undergo the fastest change. And when a gene having a new function is evolved, it is preceded by gene duplication. In this way the organism has one copy to accumulate mutations and eventually emerge as a new gene, while the other copy retains the old function required for survival.

A study of the globins reveals species divergence coincidental with duplication of genetic material in the last billion years. The alpha and beta chains of hemoglobin, presently formed from separate genes on different chromosomes, were derived from a single ancestral gene by an ancient duplication of the genetic structure. The hemoglobin of the lamprey, a primitive fish, still retains a single chain. And too, myoglobin, a related heme protein found in muscle, evidently is the result of a divergence from hemoglobin by an ancient duplication that took place in the pre-Cambrian, approximately a billion years ago. In this way, as organisms evolved to more advanced levels, their genetic material and associated proteins expanded in number and complexity.

A gene contains an average of 1,000 nucleotides; and in man the total number of nucleotides making up a genome (the set of chromosomes and all their genes) is about 4 billion. This is roughly the same for all mammals. In a study of seven proteins, Jack King and Thomas Jukes[7] found in the evolutionary history of mammals the number of amino acid substitutions averaged 2 to 2.5 per year in a species. If the average generation time in the line to man was 10 years, there would have been 20 substitutions in each generation. This is much too rapid for a stable species; variants would be lost before natural selection would have time to derive potential advantages that they might present. For this reason, the majority of changes in proteins have been neutral or almost neutral in natural selection.

The rate at which different proteins evolve depends upon the number of amino acids that are invariant for the protein to be functional. To study the rate of molecular evolution Russell Doolittle and Birger Blombäck[8] chose the fibrino-peptides. Unlike hemoglobin and cytochrome c, the fibrinopeptides do not have to conserve particular amino acids for their biological role. They are the peptide segments, 13 to 21 amino acids long, that are cut out of fibrinogen when it is converted to fibrin in a blood clot. Since fibrinopeptides can consist of virtually any amino acid, they are excellent for studying the unrestricted rate of mutation.

Application of the fibrinopeptides is essentially limited to the mammals, but their information is an outstanding contribution for hominoid evolution. Humans, chimpanzees, and gorillas all have identical fibrinopeptides, whereas there are differences in the fibrinopeptides of such closely related pairs as the cat–lion, dog–fox, donkey–horse, and the water buffalo–Cape buffalo. In studying the molecular evolution of the six hominoids by comparing their fibrinopeptides, Doolittle and his coworkers[9] discovered that, although the gorilla is identical to man and chimpanzee, he differs from the Asian apes, the orangutan and siamang. It appears that this chemical evidence of man's relationship to the African apes supports Leakey's thesis that our early progenitors lived in Africa.

Paleontologists tend to place the emergence of man distinct from the other primates as early as 14 million years ago.[10] Molecular evolutionists, on the other hand, believe that man and the African apes diverged as late as 4 to 5 million years ago.[11] For better or for worse, the chemical analysis and comparisons are less susceptible to subjectivity than taxonomy. Consequently, we learn that the molecular relationship between man and the other primates is much closer than is generally believed—and actually closer than between members of some other recognized species. Of 50 different proteins isolated from man and chimpanzee, the difference in amino acid sequences has been found to be less than 1 percent.

There are two forms of chromosomal organization: the single chromosome of bacteria and blue-green algae as a double-stranded DNA in a closed loop, and the beadlike chain of chromosomes of the higher organisms. Mutations can occur by errors in a number of ways in replication of the chromosome arrangement, but these chromosomal aberrations are almost always lethal. On the other hand, the effect of a mutation of the genes resulting in amino acid substitution depends upon the protein. If the substitution has no effect on the function of the protein, it is neutral; but if the substitution alters the shape and efficiency of the protein, it will be deleterious to the organism, and the mutation will not survive by loss of the individual from reproduction.

A species is an interbreeding pool, so any surviving mutation can spread throughout a breeding range. New species arise when mutations accumulate in reproductively isolated groups. As the changes from mutations become prevalent, interbreeding with populations of the ancestral species is first rendered unusual, then eventually it becomes impossible.

Whereas substitutions of invariant amino acids in existing proteins is a mutation that is not retained, it is the means by which new proteins achieved their optimal amino acid sequence for functionality. In the early stages of life on earth the initial proteins were probably of low efficiency, but were continually improved upon by amino acid substitution until they attained maximum effectiveness. Since then, the sequence responsible for their function has been rigorously conserved. But without mutation, life would have remained on the level of the first primitive cells on primordial earth.

It appears that life began on earth with a few basic reactions to synthesize components, and hence perpetuated itself by reproduction. The proteins for these fundamental reactions were conserved, and as organisms evolved, they developed improvements by introducing new reactions to the old ones and refining the efficiency of biochemical conversions. Because of this conservative manner in which evolution proceeds, we all have within us "relics" of our progenitors, even of the ancient microscopic organisms that floated in Archean seas. Life does not discard what the survival of the species has depended upon, but retains the old alongside any new developments. For this reason, many of the basic biochemical reactions derived from extremely remote ancestors, little changed in billions of years, can be seen in the biochemical architecture of contemporary plants and animals.

One such vestige of our origins in an anaerobic environment is a metabolic pathway for glycolysis. A main source of nutrition now and probably from the beginning is the metabolism of carbohydrates. In the simple process of fermentation, glucose is broken down to pyruvic acid, yielding 11.3 kilojoules of chemical energy per mole of sugar. Contemporary anaerobic organisms carry out this conversion in 10 reactions. Since the development of this basic series of reactions, it has been extended by some organisms to metabolizing pyruvic acid farther to lactate, propionate, acetate, ethanol, acetone, butyric acid, and higher fatty acids. Despite the fact that the decomposition of carbohydrates to pyruvic acid by fermentation has been overshadowed by more evolved metabolic pathways in all except the anaerobic organisms, it is preserved in the biochemistry

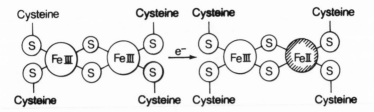

Figure 10.4. Model of the active site of ferredoxin.

of higher plants and animals as a relic from an age when it was the sole means our primitive anaerobic ancestors had of extracting chemical energy.

But the development of the whole biochemical structure has branched out from a few fundamental reactions. The result of this is that the original reactions can no longer be changed. Too much of the overall biochemistry of the organisms stems from them and depends upon their products. For example, acetate is involved in energy transfer systems, but it is also the essential starting material of such diverse components as carbohydrates, amino acids, and fats. Even if some other substance could be found that worked better than acetate, it could not be adopted without devastating consequences because so many reactions require it. A single change would pyramid into changing much of the metabolism of the organism.

This conservation of biochemical processes accounts for the close unity of all living things in the very fundamental steps that make life possible. This is why the determination of the amino acid sequences of many proteins and the application of computers have allowed biochemists to trace back the evolutionary development of today's plants and animals on a molecular level. Whereas the genetic complement of contemporary mammals is estimated to be 100,000 genes, the number of metabolic reactions that the various phyla have in common is quite small. From this it appears that all animals and man descended from an ancestral primordial cell that had only about 200 genes.[12]

The gene that transcribes cytochrome c dates back as early as the development of the eucaryotic cell 1.3 billion years ago. But a universal protein that may extend as early as the beginning of life on earth is ferredoxin, an iron-containing protein that is vital in photochemical reactions for electron transport to cellular energy storage. Ferredoxin has a reducing potential near that of molecular hydrogen, making it the most highly reducing stable compound in a cell and suggesting that it evolved at a time when the earth's atmosphere was still strongly reducing.

The various roles of ferredoxin in the cell are fundamental: it assists in the ATP formation by radiation;[13] it participates in the reduction of carbon dioxide to pyruvate; and it is used in the fixation of nitrogen.[14] Apparently ferredoxin is more ancient than nicotinamide adenine dinucleotide (NAD), a ubiquitous

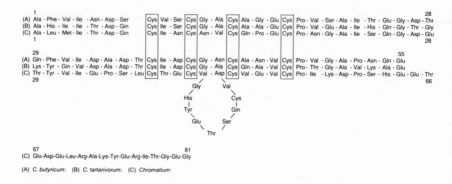

Figure 10.5. Comparison of sequences of *Clostridium* and *Chromatium* ferredoxins.

reducing agent in cells. In the primitive microbe *Clostridium pasteurianum* and in the photosynthetic bacterium *Chromatium thiosulfatophilium* ferredoxin participates directly as a reductant of carbon dioxide with acetyl coenzyme A, instead of through NAD.[15]

Ferredoxin is an iron–sulfur protein of only 55 amino acids, consisting of an unusually high proportion of the smaller and thermodynamically stable amino acids: glycine, alanine, serine, aspartic acid, and cysteine. From a study of the sequence of amino acids of ferredoxin, Richard Eck and Margaret Dayhoff[16] concluded that the protein had an ancestral sequence of 29 units and that the original molecule was based on a repeating sequence of alanine, serine, aspartic acid, and glycine. It appears that the original genetic mechanism was a sequence of 12 nucleotides that doubled, then doubled again, making a long, repetitive chain. Later, as the genetic code became more complex, other amino acids, including cysteine, were added and the sulfide bond of the cysteine became attached to the iron. Eventually, four cysteines were added by mutation and two identical chains combined to make an intricate protein–iron–sulfide complex of greatly increased efficiency.

Ferredoxin occurs in the primitive anaerobic organisms, both photosynthetic and nonphotosynthetic, and is basic to cell chemistry. The simplicity and evolutionary development of ferredoxin suggest that it may have been one of the earliest proteins formed by life on earth.[17]

Life extends uninterruptedly from the first living cell on earth down through the ages to all its living descendants. The genetic information coded in the thread of DNA nucleotides has been passed on from generation to generation, slowly being improved upon, becoming more efficient, adding new genes, and progressively moving toward the higher forms of plants and animals.

All of today's descendants are genetically the same age. But all have not climbed the evolutionary ladder to the same heights. There are the "living fossils"—plants and animals that exist little changed from their fossil ancestors.

The shark has remained essentially the same for the last 70 million years, the horseshoe crab for 180, and the cockroach, scorpion, millipede, and nautilus have changed little in several hundred million years. The ginkgo tree of China flourished during the age of the dinosaurs, and in 1958 off the coast of southwest Africa some fishermen caught in their nets a strange fish they had never seen before. And for good reason: the fish was a coelacanth, a primitive species thought by paleontologists to have become extinct around 150 million years ago.

The living fossils extend even to the most ancient forms of life. About 15 years ago, Sanford Siegel, a University of Hawaii botanist who was on a chance visit to Harlech Castle in Wales, observed tourists honoring a time-seasoned practice passed down from medieval days. In the manner of the knights of old, or at least of those of the soldiers who stood guard duty, they were urinating at the base of the castle walls. Siegel, supported by the National Aeronautics and Space Administration, had worked for several years investigating microorganisms living under harsh environments likely to be encountered in space travel. Because ammonia is a major component of the atmosphere of Jupiter and may have been common on primitive earth, it was one of the environments studied. A natural condition high in ammonia would be soil saturated with urine.

Siegel returned home with soil samples he had collected from around the castle walls at Harlech and attempted to culture them in concentrated ammonium hydroxide. Most organisms would have been killed or greatly inhibited by the medium, but Siegel observed one growing in microscopic clusters of star-shaped bodies attached to slender stalks. It did not fit a description of any known living organism, although it did closely resemble a Precambrian microfossil Barghoorn had discovered in a 2-billion-year-old Gunflint chert at Kakabeka, Ontario, Canada, which he had named *Kakabekia umbrellata*.[18]

Siegel hypothesized that he had found a living microfossil that was an obligate ammonophile, an organism that requires ammonia to grow. Since he collected the first specimens at Harlech Castle, studies by Siegel and his wife, Barbara,[19] have shown that the organism's need for ammonia is not absolute. The organism has also been found in soils from Alaska, Iceland, and various alpine regions that are low in ammonia but high in alkalinity. It does not need oxygen, but, unlike most anaerobic bacteria, is not killed by it. Siegel's discovery, which may be a living relative of Barghoorn's fossil from the Middle Proterozoic period, was named *Kakabekia barghoorniana*.

The cyanobacteria, although widely dispersed, have probably remained essentially the same for at least a billion years. And the anaerobic microbes that were dominant for the first 2 billion years of life before free oxygen existed in any significant amounts have managed to remain with us in sheltered niches and are represented today by the clostridia that survive to give us tetanus, botulism, and gas gangrene. The clostridia, which lack even cytochrome c, are listed as one of the most primitive forms of life. Even the photosynthetic bacteria, those primitive microbes that have the ability to synthesize organic matter from carbon

Figure 10.6. The gatehouse at Harlech Castle.

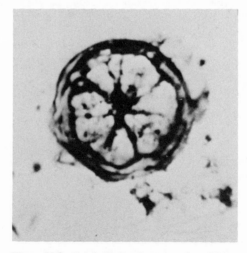

Figure 10.7. *Kakabekia barghoorniana* from Wales.

dioxide in an anaerobic environment using hydrogen sulfide as their source of hydrogen and requiring only acetic acid, light, and some minerals, survive as one of the earliest life forms on earth.

But, although the living fossils appear to be frozen in time while other plants and animals are undergoing dynamic evolution, this is not the case. Even the genes of these relics of the past have undergone mutation at the same rate as all other living descendants. In order to maintain an unchanged morphology they apparently have been under an incessant action of natural selection for hundreds of millions and even billions of years, while a steady stream of almost neutral genetic variations has flowed through, transforming their informational molecules tremendously.

All living organisms have within their biochemical structures traces of the events that have led to the advancement of life for the last 3.5 billion years. Unlike the fossil record, which has left many branches that became deadends, the chemical relics within all living things are derived from direct ancestry. Application of molecular evolution is an immensely valuable technique. With it, scientists can conceivably fulfill an old dream—trace back the evolution of the species to the very beginning of life itself.

11
Two Kinds of Life

The difference between procaryotes and eucaryotes in structural organization and versatility is enormous. On the one side are the procaryotes, of extremely diverse metabolic processes, yet essentially packets of chemicals mediating their reactions in the cytoplasm. On the other side are the eucaryotes, ten times the bacterial diameter with a volume 1,000 times larger, constructed of subcellular units bound in lipid membranes and carrying out specific biochemical operations relatively independently of the rest of the cell. Whereas bacteria reproduce by simply copying their single-looped DNA that floats in the cytoplasm and separating by fission, each half carrying with it cytoplasmic enzymes to carry on the cellular activities, the eucaryotes have genetic material a thousand times larger in a nucleus bound separate from the cytoplasm with reproduction by the elaborate mitotic mechanism.

Because of the simpler nature of the procaryotic cell, it is reasonable to assume that eucaryotes evolved from procaryotes—but evolution has left no obvious trail of stages to retrace. There is a discontinuity from one cell type to the other. They are like two different kinds of life.

The discontinuity between eucaryotes and procaryotes is exceeded only by the chasm that was bridged to bring into being the first living organisms. It was a biological leap equal to or greater than the crossover from single-celled eucaryotic organisms to the multicellular animals and man. And it took place at some time in the distant past, long before plants and animals existed, when the earth was changing in character and taking on features that we know today. What happened long ago before the Cambrian, in the world of the microbe, that led to the creation of a new form of life—the eucaryotic cell?

When the electron microscope extended the biologists' vision from a magnification of 1,000 to 100,000, it brought into focus the intricacies of microbial structure. At a magnification of 10,000 the globular bacterium staphylococcus can be seen moving with the aid of a propellerlike flagellum; at a magnification of 26,000, the bacillus of chicken tuberculosis no longer looks like a small fiber but like a fully grown worm; and at 100,000 times magnification the giant molecules of cells become just barely recognizable as fuzzy, amorphous specks. Our vision has been carried across the stages of life's smallest levels to the threshold of the world of chemical molecules.

The bacterial cell is the form of life that anchors the biological realm to the architecture that looms up out of the imperceptible smallness of atoms and their creations. The common bacteria, spherical types called cocci, are about 2 micrometers in diameter, although the rod-shapped bacilli can be up to 10 micrometers long. The smallest species are mycoplasmas, which, at 0.135 micrometers in diameter, are only 1,350 times larger than the diameter of the hydrogen atom.

The bacteria are apparently the smallest biological systems that can function as autonomous units. The degree of their smallness is difficult to imagine. There are more bacteria on and in the human body that there are people on earth. A teaspoon of soil can contain literally billions of them. And 40,000 side by side would extend only one inch. But the smallness of bacteria disguises their effectiveness. For they are deceptively sophisticated in their role in the biosphere.

The success of the bacteria is due to their extremely diverse metabolic processes. All living organisms have in common certain basic metabolic patterns, but the bacteria are unrivaled in the variety of their biochemistry. Bacterial species can derive energy by oxidizing such diverse material as ammonia, sulfur, iron, and nitrites. Some bacteria can synthesize all their cellular requirements—the complete compendium of proteins, carbohydrates, fats, and nucleic acids—from simple inorganic salts, ammonia, carbon dioxide, and water.

This versatility has allowed bacteria to expand into ecological environments far beyond the limits of other forms of life. They have been found growing in the 92–100°C (198–212°F) waters of the boiling springs of New Zealand and Iceland and in the Arctic waters at temperatures of − 12°C (10°F). Bacteria and algae live in the Great Salt Lake and the Dead Sea, where the salt content is 29 percent. And they exist in bogs, springs, and lakes composed of strong acids.

Bacteria are bracketed into their small size range by selective pressures stemming from their level of life. All organisms take in food to grow and expel metabolic wastes. On the microbial level these materials must pass through the cellular membrane. Since life processes are confined to a limited volume, the rate of these reactions depends on how rapidly substances can be absorbed and expelled by the cell. Because the proportionality of surface to volume increases as the size decreases, there is a clear, distinctive advantage to smallness. The surface to volume ratio of man is about 20, but with bacteria it soars to 9 million. This is why bacteria, yeasts, and other microbes have a tremendous rate of metabolism. The cell of *E. coli* added to a thimble of nutrient broth can increase a thousandfold in three hours; in a day it can spawn billions of progeny; and if the rate could be sustained, in 30 days it would engulf the earth.

At the lower end of their size range the bacteria are held to a volume large enough to contain the necessary number of molecules to be a functional cell. As their size increases, expansion becomes restricted by the decreasing metabolic rate, but in particular by the limitations of their energy supply. With a few exceptions, bacteria are fermenters. They break down substances for their chemical energy. Fermentation, however, is a low-yield energy source. And for those organisms which rely upon it as the sole means to fuel their processes, they are forever condemned to a low level of existence.

Just as there are advantages to being small, there are even greater advantages to being large if the organism can muster an energy concentration to sustain the increased size and organizational complexity. And when the conditions were favorable for the existence of a larger cellular life form, it evolved to take its place in the living world. For 2 billion years the procaryotes were the only form of life, but during that long interval they created the conditions that allowed for life to step up to a larger and more complex organization. In their diversity they created a wide variety of enzymic pathways among themselves, and the blue-green algae freed into the environment sufficient oxygen to make possible the complete oxidation of glucose to carbon dioxide and water through respiration. The biologists of the last century recognized that there were two cellular types of life, but the evolutionary relationship between them remained obscured. How the larger and more complex stage of life was achieved through evolution was purely speculative until only recently.

The microbiologists of the nineteenth century could see that eucaryotic cells differed from the smaller bacterial cells by size and membrane-bound organelles, but were held back from seeing the structural detail that the electron microscope later revealed. They could, however, study the stained and pigmented subcellular components. And A. F. W. Schimper a hundred years ago was one of the botanists to discover that chloroplasts are not synthesized *de novo* but are produced by division.[1] He seems to have been the first to notice their similarity to the free-living blue-green algae. Like the chloroplasts, the mitochondria also reproduce by division, and in 1890, Richard Altmann,[2] a professor of anatomy at Leipzig, hypothesized that these two particles were comparable to bacteria and were actually symbionts. The difference in the two types of cells is so great that procaryotes tend to be no larger than the subcellular units of the eucaryotes. The idea that mitochondria and chloroplasts are bacterial cells inside one another was received, however, with much criticism at the time and was eventually forgotten.

Twenty years later, K. C. Mereschowsky[3] in Russia developed the concept of symbiotic origin of organelles, but his ideas for the most part remained unknown to scientists outside of Russia because of the language barrier. In the 1920s, the American biologist Ivan Wallin[4] at the University of Colorado Medical

School argued forcefully for the notion that the little particles inside eucaryotic cells were symbionts of bacterial origin. The concept drew little attention until 1962, when Hans Ris and Walter Plaut[5] at the University of Wisconsin reported finding DNA in chloroplasts. Then two years later DNA was discovered in the mitochondria from both plants and animals.[6] Suddenly, a new and profound feature of eucaryotes was recognized. Unlike the procaryotes, which have only one DNA molecule, the eucaryotes actually have several genetic centers. They are polygenomic.

Not only was DNA present in chloroplasts and mitochondria, it was in a closed loop like the DNA of bacteria. This strongly suggested that these subcellular particles had self-replicating properties, as well as protein-synthesizing capabilities. Whereas procaryotes remain as individual protein-synthesizing units, eucaryotes were found to be composed of a number of separate units that synthesize their own proteins. On further examination, it was seen that the blue-green algae, the mitochondria, and the chloroplasts all have similar lamellar structures independent of the internal membrane of the cell, all have DNA, enzymes, carotenoids, cytochromes, and other components in common for their activities, all divide, and all can mutate and evolve.

The evidence was strongly indicative that eucaryotes did not evolve from procaryotes by a simple dilation into complexity through enlargement and mutation of their genetic material. Rather, something happened, either abruptly or over a long transition, that brought together diverse elements of the procaryotic world into a lasting union so complementary that it created a new form of life.

The living world of the Early and Middle Proterozoic was structured along lines of selective processes as it is today. The procaryotic organisms honed their efficiency through mutation, evolved diverse types to expand to the limits of their physical tolerance, and strengthened their gains through specialization. The expansion was not confined to the fringes of the physical environment. Like all stages of the biosphere, the procaryotic world was crisscrossed with morphological and nutritional dependencies. Only photo- and chemoautotrophs produced their organic requirements solely from inorganic substances. All others depended on predation or the metabolic waste of others for their food. Predation, parasitism, and symbiosis developed extensively throughout the biosphere.

When two or more organisms have a highly specific association based on mutual dependency it is called symbiosis, and the bonding can be anything from a casual relationship to an intimate, obligate union. The relationship usually occurs in nature under conditions that are too extreme for the range of the enzymic repertoire of any single organism. A common example is the lichen, which lives in places too arid for algae, too deficient in organic nutrients for fungi. As a result, the crusty lichen thrives as a plantlike composite where millions of cells of algae are woven in the matrix formed by the filament of a fungus, and between the symbionts an exchange of nutrients flows.

Organisms of all sizes and all kingdoms enter into symbiotic relationships.

Bacteria live in the intestinal tracts of animals, fungi cohabitate with algae, and nearly every starfish and sea urchin has one or more microbial symbionts. Between plants and bacteria nitrogen fixation is performed in nodulated legume roots, a process that neither partner is capable of carrying out alone. Symbiosis among bacteria is common, especially in anaerobic environments. In the absence of oxygen organic substances degrade slowly, so that the total breakdown of organic matter to gaseous products often requires more than one bacterial species to provide the complete enzymic degradation.

If the symbiotic theory is the explanation for the discontinuity between the two kinds of life, its confirmation must lie in chemical and biological evidence linking these parts of the eucaryotic cell believed to be of symbiotic origin to their previous existence as free-living organisms. The organelles that were symbionts would have been at one time free-living organisms and would have had as a minimum a DNA, a messenger RNA, a protein-synthesizing system, including the transfer RNAs with their enzymes and their ribosomes with their proteins and nucleic acids, a source of ATP, and a cell membrane-synthesizing system. After entering a host cell by a symbiont, redundancies between the two would have been selected against. The intracellular symbiont may therefore have lost all or none of its mechanics for independent existence by relegating indispensable metabolic functions to the host. An examination of the genetic material of contemporary organelles should yield clues to what qualities were conferred on the host and lead us back to the nature of the ancient symbiont.

Moreover, because of the conservative behavior of biological evolution, examples of naturally occurring, free-living counterparts may still be found in nature, codescendants of the symbionts with the same genetic and physiological characteristics which have survived in a changing world.

The most apparent candidate for an organelle of symbiotic origin is the chloroplast of green algae and plants—it looks like a photosynthetic procaryote. And there are many cases where nonphotosynthetic organisms exploit the chloroplasts of other things. Some organisms take in chloroplasts from algae to supplement their energy production, then eject or digest them when they are no longer needed. The giant clam is fed partly by photosynthesizing algae living between cells of its mantle. And certain sea slugs acquire chloroplasts by feeding on algae which then live independently in the slug, photosynthesizing for both themselves and their host.[7] Among the microbes is the familiar photosynthetic flagellate *Euglena*. The euglena has chloroplasts which are of value during illumination but are not essential for life. When the organism is cultivated in the presence of penicillin, it no longer produces chloroplasts and becomes indistinguishable from the common protozoan *Astasia*.[8]

The green algae were probably the first photosynthetic eucaryotes to develop with this organelle around 1 billion years ago, when progeny of the ancestral eucaryote acquired a photosynthetic symbiont. The other modern algae and higher plants then evolved from them, while another line from the early eucaryote led to the protozoa, fungi, and higher animals.

Chloroplasts have always shown a persistent individuality, probably because they were acquired as symbionts relatively recently. They have their own DNA, messenger RNA, a protein-synthesizing system, and ribosomes that are sensitive to antibiotics that affect bacteria. The chloroplasts of the green alga *Chlamydomonas* have coding for several hundred proteins, in contrast to the mitochondria, which have about one-tenth of that capacity.[9] Nevertheless, the chloroplasts have transferred some of their genetic information to the chromosomes of the host since entering into the association, because they now need the participation of both the nuclear and chloroplast genes to reproduce.

If the chloroplasts were acquired through symbiosis, they should still carry tracks of their procaryotic ancestry. Ford Doolittle[10] of Dalhousie University in Halifax, Nova Scotia, compared the similarity of ribosomal RNAs from the photosynthetic organelle of the red alga *Porphyridium* to that of its own cytoplasm, and to the same RNA from the blue-green algae and the chloroplasts of *Euglena*. He found that the organelle was more closely related to the other organisms than to the red alga itself, of which it was a part. The blue-green algae and chloroxybacteria are free-living counterparts of the chloroplasts, and the ancestor of the organelle probably resembled these photosynthetic organisms.

Organelles that seem more ancient than the chloroplasts are the mitochondria, since no algae or plants are without them. If mitochondria are descendant from a symbiont, the ancestor must have been a nonphotosynthetic microbe that was entirely dependent on oxygen for metabolism. Other bacteria carried out fermentation, but the antecedent of the mitochondria had evolved as an aerobe that thrived on the waste products from carbohydrate breakdown by other organisms. Presumably it originally found these substances in the environment, but eventually entered into a symbiotic relationship with a host cell to feed on its food-rich interior. While the small aerobe lived in the rich culture of the host's fermentation and received protection, it supplied its host with the excess energetic products from its metabolism. Natural selection against redundancy between the host and aerobe eventually led to mutual dependence. Mitochondria can no longer exist independently outside of eucaryotic cells, nor can the cells survive when deprived of their mitochondria.

Mitochondria have retained their own DNA, messenger RNA, transfer RNAs, and ribosomes. The DNA has a molecular weight of 10 to 100 million atomic mass units, but this is not enough to code for all the proteins in mitochondria.[11] The mitochondria, therefore, have lost much of their genetic control to the nuclear genome and have become obligate symbionts, depending upon products synthesized by the host cell.

If mitochondria are of bacterial origin, they too should resemble some free-living bacteria found among the microbes today. A contemporary organism that may be a codescendant with the mitochondria from the same ancestor is *Paracoccus denitrificans,* an aerobic, rod-shaped bacterium that oxidizes its fermentation products completely to carbon dioxide and water.[12] When the respiratory system of this aerobe was compared to that of mitochondria of animals and yeast, the similarity was found to be remarkable. The quinones and cytochromes from their electron transport systems are closely akin. The bacterium, however, differs by having a cell wall, by being able to reduce nitrate to nitrite, and by lacking an ATP transport system. The cell walls and nitrate reduction are superfluous to an intracellular symbiont and would have been selected against. The ATP transport system, on the other hand, would not have been present in the free aerobe but is a feature necessary for mitochondria, so this system presumably evolved with the symbiosis.

What kind of organism would have absorbed a free-living aerobe into a symbiotic association? Since the cytoplasm can be considered the interior of the host cell, it would appear that the host cell had been an anaerobe with the Embden–Meyerhof pathway for the fermentation of carbohydrates. The cytoplasmic enzymes are coded in the nuclear DNA, so it follows that the nucleus also is derived from the host cell. There are three prominent features about the nucleus of eucaryotic cells that distinguish it from the procaryotic DNA: the genetic material is arranged in chromosomes, the nucleus is bound by a lipid membrane, and replication of the nucleus is by the mitotic process. Any or all of these features may offer insights into the nature of the ancestral host cell.

The eucaryotic genetic material differs also by being complexed with basic proteins. Physically, the DNA molecule resists packing because the phosphate groups tend to be negatively charged and repel each other. The nuclear membrane allows this repulsion to be neutralized by maintaining a higher concentration of the positively charged sodium ions in the nuclear fluid than is in the cytoplasm. The histones of the genetic material also neutralize the phosphate groups and cause close packing.

Dennis Searcy and his coworkers[13] at the University of Massachusetts suggested that the heat- and acid-tolerant mycoplasma *Thermoplasma acidophilum* may be a codescendant of the eucaryotic cell. This wall-less bacterium has the Embden–Meyerhof pathway, the sterol requirements of eucaryotes, the arginine and lysine-rich, histone-like proteins to protect its DNA from heat and acid, and it enters into associations easily. When, however, the nucleotide sequence of the 16S ribosomal RNA of *T. acidophilum* was compared to the same RNA from other organisms, the analysis indicated that the bacterium was more closely related to the archaebacteria methanogens and halobacteria than to the contemporary eucaryotes.[14] Apparently the host cell came from some other type of predecessor or by another manner of development.

One consideration is that the nucleus itself may have begun as a cell inside the host cell. The DNA of the nucleus carries the code for the cytoplasmic enzymes and some of the genetic information of the organelles of symbiotic origin. Nevertheless, it is bound in its own separate membrane, suggesting that the nucleus also is a remnant of a symbiont. This idea was expressed by the Russian biologist K. C. Mereschowsky at the turn of the century and is supported by J. Goksør of Norway, J. Pickett-Heaps at the University of Colorado, and H. Hartman at the Massachusetts Institute of Technology. If the nucleus was a symbiont, however, the question is still open on what happened to the host's DNA and how the two systems were integrated.

On the other hand, if the nucleus evolved from the procaryotic DNA of the cytoplasmic host cell, then at some time in its origin the DNA acquired its own membrane. When procaryotic cells divide, their DNA and its copy attach to the growing membrane, and each goes with its respective half of the cell when it divides. Conceivably, the DNA in some ancestral bacterium attached to the membrane and became encapsulated inside the cell by a segment of the cellular membrane.

Encapsulation of the DNA inside the host cell could have resulted in a rapid introduction of the chromosome arrangement. Replication of the DNA isolated by the nuclear membrane during the early stages of the emerging eucaryotic cell before the nucleus was coordinated with the other cellular activities possibly created the potential for multiple DNAs and an abrupt expansion of the genetic material.

There are two seemingly unrelated features of eucaryotic cells that may have had a common origin. These features are the mitotic process and the apparatus for motility. Both are carried out in eucaryotes with bundles of fibers called microtubules. If Lynn Margulis[15] of Boston University is correct, both the mitotic process and the eucaryotic means of locomotion resulted from a symbiosis between the pre-eucaryotic cell and a type of bacterium called spirochetes.

In order to propel themselves, some bacteria are equipped with one or more filamentous appendages consisting of fibrous proteins. These flagella give locomotion, not by a lashing motion, but by a rhythmic contraction that moves helically through the flagellum from one end to the other. In this way motile bacteria like the *E. coli* can attain speeds of 25 micrometers per second, or about 0.0001 mile per hour.[16]

Eucaryotic cells also use hairlike cell projections for movement. When the projections are long and few, like the tail of the spermatozoa, they are called flagella and the organisms are referred to as flagellates. And when the projections are numerous and short, they are called cilia, and the organisms ciliates. Flagella and cilia are made up of bundles of fibers with a strikingly uniform structure. They are about 0.25 micrometer in diameter and consist of a circle of pairs of minute cyclinders with another pair of microtubules in the middle, giving them a 9 + 2 structural array. Because the flagella and cilia of eucaryotes are structurally dissimilar to the smaller, single-stranded bacterial flagella, they are classed together under the name undulipodia.

Among the bacteria is a cellular type where the organism is spirally twisted.

Some, the vibrios, are only slightly curved like a comma, but the spirochetes are markedly twisted. Their bodies consist of a protoplasmic cylinder bound by a plasma membrane. Long, slender, and flexious, the spirochetes are highly motile, moving through the medium with an undulating wave motion.

A striking similarity in size and movement exists between the spirochetes and the undulipodia of eucaryotic cells. And a common behavior of spirochetes is to attach to things and beat in unison. Margulis gives a notable example of spirochete behavior with a contemporary organism that may resemble some ancient occurrence that took place in the history of the eucaryote. There lives in the hindgut of the Australian termite a eucaryotic organism called *Mixotricha paradoxa* that has four undulipodia. But the *Mixotricha* does not use these appendages for locomotion. Rather, the microbe is propelled about by a half-million attached spirochetes moving in coordinated undulations, while it uses the undulipodia as rudders.[17]

The similarities between the spirochetes and the undulipodia of eucaryotic cells suggest that the latter resulted from some ancient symbiosis of spirochetelike organisms with the host cell. Apparently spirochetes attached themselves to the host cell to feed on food leaking through the membrane. In turn the symbionts conferred motility on the pre-eucaryote, a property that would have given the developing eucaryote a large advantage in seeking out food. The symbiotic theory also explains why undulipodia are able sometimes to become detached from the cell and swim away.

The chemical evidence relating the origin of the undulipodia to ancestral spirochetes is still scanty. But this may be because their symbiotic association began so much earlier than that of the mitochondria and chloroplasts. Nevertheless, RNA has been found in the kinetosomes, the small cellular bodies from which the undulipodia grow. Except for the RNA, undulipodia seem no longer to contain genetic material, their reproductive and metabolic capabilities having been passed on to the nuclear chromosomes during a long symbiotic dependency.

But structurally the undulipodia have the 9 + 2 microtubular array observed in the composition of some spirochetes, especially those found in termites. Moreover, the size of the eucaryotic appendages and the spiral bacteria is about the same. If the spirochete is a codescendant, then the symbiont during the course of evolution of the eucaryotic cell lost its plasma membrane.

A serious requirement for the evolving eucaryote would have been an orderly replication and separation of its growing DNA. Eucaryotic DNA has become a thousand times larger than the simple DNA of bacteria. To duplicate and divide this complex arrangement the eucaryotes evolved the mitotic process. On cue minute, dotlike bodies called centrioles go to opposing parts of the cell and send out spindles to attach to the chromosomes and reel them to each half of the cell. The electron microscope image of a cross section of the centriole shows that this organelle has a 9 + 0 array of microtubules. The pattern differs from the undulipodia only in the absence of the pair of cylinders in the center.

Margulis[18] suggests that the highly efficient and elaborate mitotic mechanism was acquired symbiotically from a 9 + 2 motile organism. Among the same motile spirochetes that were ancestors of the flagella of eucaryotic cells may have been symbionts that entered the cell through the membrane, never to return. They remained to play a major role by furnishing the parts for a mechanism that evolved into the elegant mitotic procedure of today.

The discovery of the polygenomic nature of the eucaryotic cell led biologists to realize that the transition from procaryotes to a higher stage of life was not by evolution from a single cell type—but from a consortium. In our Precambrian past we have an ancestor that resulted from the merging of several phylogenetic ancestral lines. And the manner in which this association between diverse types of organisms took place is a common feature that continues to occur in the biological world today. Yet the symbiosis that created the eucaryotic cell was of unparalleled significance, for it crossed a threshold that opened up in fairly rapid succession to all higher forms of life, including mankind.

Evolution of the eucaryotic cell was spectacularly successful and led to our being. But life had been confined to the microbial level for 80 percent of its existence before plants and animals emerged. Why did it take so long?

The earth 2 billion years ago was a world in transition. Ever since the blue-green algae evolved with an enzymic capacity to photocatalyze the decomposition of water, they had been releasing oxygen into the environment, only to have it rapidly absorbed by the ferrous ions of the Archean seas and precipitated as insoluble oxides with silica as banded iron deposits. But after a billion years the "oxygen sink" was becoming depleted of the reduced ions. Then in a last surge of deposition about 1.8 billion years ago the seas were swept of the soluble iron and molecular oxygen began to accumulate in the atmosphere and oceans.[19] The reduced conditions of the ancient order were put behind it as the earth moved toward an oxygenated state with its modern characteristics; and as the oxygen level grew, it carried with it the adaptation of organisms to its powerful influence.

To the anaerobes oxygen was a deadly poison to be protected against or to escape from in sheltered niches. To the aerobes oxygen was the promise of oxidative phosphorylation and a concentration of energy 18 times greater than the feeble output of fermentation.

The eucaryotes seem to have appeared sometime around 1.4 billion years ago, and with the thrust of total glycolysis they had the potential to propel themselves into a whole new world. Within 600 to 700 million years the coelenterates, jellyfish, and sponges appeared in the seas across the hurdle of multicellularity. And as the avenues of the new dimension opened, they were filled rapidly within 100 million years by the new life forms in an explosive expansion into all basic anatomic types. Only the colonization of the continents and the vertebrates were to come later.

There is a consensus among earth scientists that the vast amount of free oxygen in the atmosphere is an accumulation over the eons of oxygen released into the environment by photosynthetic organisms. This process continues today through plants and photosynthetic microorganisms, with an estimated 90 percent of the oxygen being generated by the phytoplankton of the oceans. The atmospheric oxygen at its present 21 percent concentration does more than provide conditions conducive for aerobic life; it provides an ozone shield in the upper atmosphere that screens out the high-energy ultraviolet radiation in the 240–290 nanometer range that is destructive to organic substances.

Ultraviolet light has a profound effect on living matter. It destroys unprotected microorganisms rapidly by deactivating DNA, and it is the cause of sunburn. Today, practically all the ultraviolet radiation is screened out by a layer of ozone in the upper reaches of the atmosphere that is generated by the action of shortwave ultraviolet light on oxygen.* But before the presence of atmospheric oxygen the ultraviolet rays of sunlight would have shone down on the earth's surface and into the upper layers of the oceans with a fierce intensity, wreaking havoc on the fragile nature of unsheltered life.

The earth between 1.8 billion years ago and the dawning of the Cambrian period 570 million years ago underwent change both physically and biologically as the level of atmospheric oxygen grew. When free oxygen entered the environment there were two biological responses: aerobes employing a restricted, low-demand, oxidative metabolism commenced; and the anaerobes sought means to survive its deadly effects. These reactions would have happened simultaneously. When the partial pressure of oxygen is high enough for oxidative phosphorylation, hydrogenase and similar anaerobic catalytic mechanisms are oxidized into inactivity. This level of oxygen where fermentation is suppressed and aerobiosis is possible was first observed by Pasteur and is about 1 percent of today's atmospheric concentration.[20]

Until the accumulation of an adequate ozone shield for ultraviolet light, the habitat for photosynthesizers would have been confined to a photic zone screened by a layer of water or agglutinized sediments. As a growing ozone layer diminished the ultraviolet radiation, phytoplankton would have expanded to the surface layers where photosynthesis is more active. Occupation of the open seas by phytoplankton most likely accelerated the production of hydrospheric and atmospheric oxygen immensely, in turn increasing the ozone layer and expanding the life zone.

Although the order of events can be outlined, setting them to a definite time scale is more difficult. Establishing the time of biological response and the

---

* At the pressure of one atmosphere the amount of ozone in this protective shield would be a layer of pure gas only 3 millimeters, or one-eighth of an inch, thick. It is diluted, of course, with the other air gases and is at an altitude of 10 to 40 miles, so that the ozone-containing layer is actually several miles deep. Nevertheless, if the amount of ozone in the atmosphere were reduced to one-third, our skins would be destroyed on exposure in a matter of minutes. On the other hand, if the amount were doubled, the human race would probably die out for lack of an essential vitamin, while being smothered by an accumulating mass of bacteria.

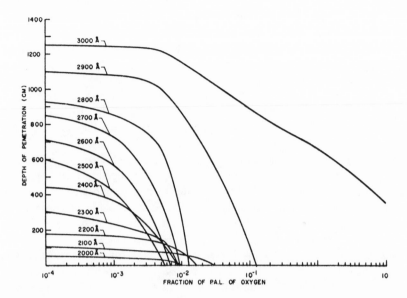

Figure 11.1. Penetration of ultraviolet radiation in water at various combinations of oxygen and ozone atmospheres.

appearance of the ozone layer depends on knowing the levels of atmospheric oxygen during the pre-Cambrian, and this can only be inferred indirectly. We do know from the fossil record that animals with hard parts evolved in an explosive manner in the seas at the beginning of the Cambrian period and colonized the continents about a hundred million years later.

Without atmospheric oxygen, ultraviolet light at the 260-nanometer wavelength would have penetrated water to a depth of about 7 meters. L. W. Berkner and L. C. Marshall[21] have shown that this biologically damaging radiation would have been abruptly reduced at an oxygen level between 1 and 10 percent of the Pasteur point, thus allowing the phytoplankton to expand throughout the surface layers of the oceans and rapidly increasing the rate of oxygen buildup. When the oxygen level reached about 3 percent of the Pasteur point, metabolism of relatively low oxygen requirements could have commenced with a wave of multicellularization appearing among the eucaryotes.

Preston Cloud has pointed out that a slow increase of oxygen after 1.8 billion years ago was accompanied by increasing prevalence of limestone and dolomite deposits, indicating a decreasing level of atmospheric carbon dioxide. The large increase of oxygen and decrease of carbon dioxide toward the end of the pre-Cambrian period may have been reflected in the widespread glacial deposits that occurred at that time. The paucity of oxygen during the pre-Cambrian is supported by the rarity of glauconite (a potassium iron silicate) and sedimentary calcium sulfate deposits during that time, in contrast to their abundance in younger sediments.

The appearance of multicellular animals doubtless was preceded by a long evolution of underlying cellular processes that made their sudden appearance possible. There are biological indications that oxygen was an influence in the environment much earlier than the geological record suggests. Margulis and Rambler[22] have reported that procaryotes, even closely related species, often differ markedly in their response to oxygen, suggesting that they evolved their major metabolic patterns in response to varying and rising oxygen levels. Since eucaryotes are extremely uniform in their response to oxygen, this gives the appearance that mechanics to handle oxygen were developed in various procaryotic groups and were established by the time eucaryotes appeared.

This seems to be supported by the apparent antiquity of oxidases to deactivate peroxides. Molecular sequencing has shown a high degree of similarity in the N-terminal sequences of superoxide dismutase between Fe- and Mn-enzymes of *E. coli,* mitochondrial Mn-enzyme, and *Bacillus stearothermophilus,* which makes it likely that the ancestor at the base of the *Bacillus* branch had a functional superoxide dismutase.[23] This, however, does not necessarily indicate a general level of atmospheric and dissolved oxygen in the environment during the Archean. Procaryotic types in close association with oxygen-releasing blue-green algae could have evolved oxidases and oxidative metabolism long before the atmosphere reached an oxidized state, much in the way methanogens today draw on gaseous hydrogen liberated by clostridia.

Margulis presents a persuasive case that microorganisms could have grown on the edges of continents long before the advance of plants and animals during the Cambrian. An atmospheric oxygen level only 1 percent of the Pasteur point would have allowed sufficient ozone to shield the earth's surface from most lethal radiation. Moreover, procaryotic microorganisms have many means to protect themselves against radiation, including an enzyme-mediated repair mechanism for damaged DNA. The continents, therefore, may have been invaded by life from the seas even so long ago as the Archean aeon. Extending from the coasts would have been a sheath of blue-green and purple scum of photosynthetic bacteria. Beneath this mat layers of yellow, brown, and black anaerobic and nonphotosynthetic bacteria eked out an existence, while other bacteria with fruiting bodies and fungal types grew between particles of soil.[24]

The long time it took eucaryotes to evolve to complex multicellular animals of the Cambrian is believed by Margulis to be due to biological, rather than environmental, factors. The advance to the complex development of creatures with shell and bone required the elaborate processes of mitosis-meiosis and calcium regulation. Mitosis became possible only through evolution of a cell cycle, chromosomal organization, and a microtubule-mediated gene distribution mechanism. Mitotic microtubule systems require low and regulated calcium ion concentrations for protein polymerization and may have preadapted organisms for the later evolution of protective and supportive calcium carbonate features. All of these processes require oxygen and would have taken a long time to develop.

When microorganisms are exposed to an oxygen gradient, either in the laboratory or in nature as in soil depths, their distribution ends up reflecting the level of oxygen. Procaryotes align themselves in the gradient by their varying oxygen requirements, showing that this gas is still a selective agent with them. Eucaryotes, on the other hand, as a group tend to associate with the fully aerobic level. Margulis feels that this is a strong indication that the earth's atmosphere was fully oxygenated when eucaryotes evolved.[25]

Evaluating the rate of evolution, however, is as uncertain as estimating oxygen concentrations of the pre-Cambrian atmosphere. Biological evolution is not a process in isolation, but one bound closely to the physical environment. Oxygenation of the environment by photosynthetic organisms created an energy potential upon whose force the eucaryotes drew in order to evolve. It seems unlikely that an appreciable time lag would develop between biological and geological evolution. A rapid acceleration of the oxygen level toward the end of the pre-Cambrian probably sparked a comparable acceleration in biological evolution, much in the way that newly opened ecological niches set off an explosive diversification to occupy them.

12
Archaebacteria

The ascent of biological systems from an assemblage of simple chemicals nearly 4 billion years ago toward greater size and complexity has not been an even advance. The adaptation mechanism of Darwinian evolution explains the means of progressing to the most efficient organism within an ecological niche, but on two major occasions biological creations rose out of the established level into a larger expanse of the physical world—into a whole new arena of openly exposed ecological niches. One such occasion was the formation of the eucaryotic cell, which was several orders of magnitude larger than the bacterial cell. Another occasion was when life catapulted out of the world of the microbe into multicellularity—the physical dimension of plants and animals.

On all levels the tenets of Darwinism remain valid: organisms overproduce their kind, and competition among individuals improves the adaptation. But the competitiveness of evolution in the transition from microbe to man has shifted from the chemical to the physical. Life on the bacterial level is competition by biochemical processes to consume those substances of chemical energy. The advance into eucaryotic and multicellular size threw the competition to the claw and the jaw, to physical shapes and appendages to wrest from within the biosphere sources of organic substances. Only the photosynthesizers draw their energy from the outside, while all others prey upon the biologial composition of the biosphere itself for their survival.

The progression from biochemical diversity to adaptation of structural detail has affected the work of the paleontologist whose task is to reconstruct the evolutionary scale. The fossil record, which reveals morphologies, was a valuable means of drawing the phylogenetic lines of plants and animals since their appearance at the dawning of the Cambrian period 570 million years ago. Reconstructing the phylogenetic lineage of the bacteria, on the other hand, has been a different matter. The diversity of bacteria is primarily biochemical—the breadth of their expansion was with different enzymic processes to tap the energy sealed in the many types of chemical substances, while their physical shapes varied little and offer few clues to their evolutionary order. As a result, the very simplicity of bacterial morphologies prevents their microfossils from being particularly informative.

In order to construct the relationship between the many kinds of bacteria it has been necessary to descend to another level of complexity—the chemical realm— where molecules grow in size out of the imperceptibly small to become the giant molecules of life processes. When the DNA molecule undergoes mutational modification of its nucleotide sequence, both the nucleic acids and the proteins reflect the change in their compositions. Because of this, these biopolymers become molecular chronometers that can be used for establishing the time when species diverged from a common ancestor. By comparing homologous molecules from various organisms, the biologist can, therefore, set to order the genealogical relations of all living things, including the bewildering assortment of bacteria.

The phylogenetic history of the bacteria can give us a perspective of those biochemical features that are the most ancient. And once the order is recognized, we can then ask with some assurance of arriving at an answer: What was the earliest form of life?

In the chemical microcosm of the bacteria the solution to the riddle of phylogenetic order lies in reconstructing the descent of the metabolic processes. This can be achieved by comparing the chronometric molecules—the nucleic acids and proteins—involved in the biochemical pathways, with the choice focused on those biochemicals whose existence extends back as close to the ancestral root of life as possible. Both the ferredoxins and the cytochromes are associated with the electron transport system in energy utilization and are clearly basic. And since any translation of genetic information to proteins is carried out on ribosomes, the ribosomal molecules have an ancestry that should extend to the time of appearance of the bacterial form of life.

Margaret Dayhoff and Robert Schwartz[1] at the Georgetown University Medical School in Washington, D.C., have presented a phylogenetic tree of bacterial ancestry based on the comparative sequences of bacterial ferredoxin, 5S ribosomal RNA, and cytochrome c. They found three zones of metabolic innovation that appeared in the pre-Cambrian. The oldest is the anaerobic stem. The middle zone reflects the rise of oxygen-liberating photosynthesis and aerobic respiration. From this area evolved the aerobic bacteria and the symbiotic association that led to the eucaryotes.

Based on ferredoxin sequences, the species of anaerobic bacteria that seems closest to the root of the ancestral origins is the clostridia. Ferredoxins are used

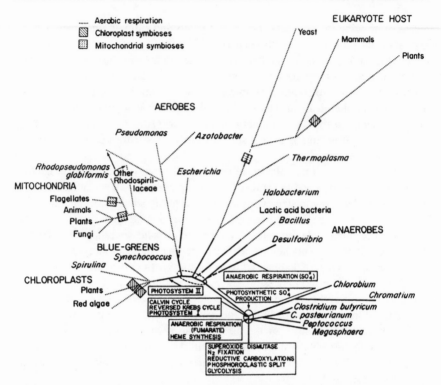

Figure 12.1. Phylogenetic tree of procaryotes and eucaryotes based on ferredoxin and cytochrome c sequences.

by anaerobes in the splitting of pyruvic acid to yield acetyle phosphate and in the reductive carboxylation of an acyl coenzyme A derivative. The ferredoxin, therefore, allowed the ancient clostridia to meet their needs for ATP and reducing power for biosynthetic reactions by glycolysis and the splitting of pyruvate. Some clostridia have the ability to carry out the total synthesis of acetate from carbon dioxide during fermentation, while others reduce fumarate to acetate. Thus, it can be inferred that the early clostridial species were able to reduce carbon dioxide, synthesize heme for their cytochromes, and use fumarate as a terminal electron acceptor for their anaerobic energy extraction procedure.

Nitrogen fixation is also a process that occurs in many species of *Clostridium*, as well as in the green sulfur bacterium *Chlorobium* and in most of the sulfate reducing *Desulfovibrio*. The reduction of nitrogen to ammonia requires four ATP molecules per electron pair and six electrons from a low-potential reductant such as ferredoxin. The electron carriers are a Fe-protein and MoFe-protein that are unrelated to ferredoxin but also contain iron–sulfur clusters. In many cases a nitrogenase component from one bacterial species can complement the other component of another species, suggesting that the function of proteins in nitrogen fixation has been conserved and that this mechanism is a primitive feature among life processes.[2]

Anoxygenic photosynthesis existed with early organisms long before the oxygen-liberating photocatalytic dissociation of water appeared. The Chlorobiaceae (green sulfur bacteria) and the Chromatiaceae (purple sulfur bacteria) are photosynthesizing organisms of ancient ancestry that use the sulfur of hydrogen sulfide rather than the oxygen of water as their source of electrons. In *Chlorobium* light is absorbed by bacteriochlorophyll, which is used to reduce NAD* by a flow of electrons from the initial electron acceptor via ferredoxin and NAD reductase. Sulfur then replenishes the electrons to the bacteriochlorophyll through a cytochrome chain. Chromatiaceae, on the other hand, generate ATP from a cyclic electron transport system which passes the electron from the primary acceptor to ubiquinone and a sequence of cytochromes back to the bacteriochlorophyll. This photosynthetic process, designated photosystem I, includes cyclic, noncyclic, and reverse electron-delivery systems, all early innovations; and organisms of this type presumably evolved soon after the clostridia.

Whereas Chlorobiaceae and Chromatiaceae produce sulfate by extracting electrons from sulfur, the *Desulfovibrio* derives its energy by reducing sulfate. The desulfovibrios retained many properties of the anaerobic stem like the clostridia but are unique in the manner in which they reduce sulfate. The process involves adenosine 5-phosphosulfate and a specific carrier protein called cytochrome c_3. It therefore has been suggested that sulfate-reducing *Desulfovibrio* evolved after the sulfate-producing photosynthesizers.[3] Manfried Schidlowski of the Max Planck Institute in West Germany obtained isotopic data on samples from the Aldan Shield of Siberia and banded iron formations from Canada that suggest that sulfate reducers appeared between 2.8 and 3 billion years ago.[4]

Sometime thereafter and before 2 billion years ago the blue-green algae arose from the anaerobic stem using water as an electron donor for the photodissimilation of carbon dioxide. This oxygen-releasing photosynthesis employs chlorophyll a and phycobiliproteins as light-absorbing pigments, and, unlike other photosynthetic bacteria, which use NADH as the reductant of carbon dioxide, the blue-green algae use NADPH. The immense significance in the appearance of the blue-green algae in evolution is that they released free oxygen into the environment, eventually changing the chemical character of the oceans and atmosphere and creating the oxygenated conditions for aerobic respiration.

The forced adaptation to the presence of oxygen led organisms to evolve an electron transport system that terminated on oxygen. The phylogenetic tree indicates that three groups of facultative anaerobes—*Bacillus,* lactic acid bacteria, and *Escherishia coli*—diverged from the trunk about the same time as the blue-green algae by adjusting to the presence of oxygen. In this adaptation, the bacilli evolved a cytochrome-dependent electron transport system, and the lactic acid

* Nicotinamide adenine dinucleotide.

bacteria, a flavin-terminating system. The *E. coli,* on the other hand, has two oxygen-terminating electron transport pathways that depend upon the level of oxygen tension. At conditions of high aeration the pathway terminates on cytochrome c; but at lower tensions it terminates on cytochrome d. These occurrences of various unrelated terminal components of aerobic respiration suggest that oxygen adaptation evolved independently on several separate occasions.

Once aerobic respiration and the Krebs cycle for total glycolysis became established in the procaryotic stem, the large energy bonus made possible the larger and more complex eucaryotic cell. The mitotic process, the biosynthesis of steroids, polyunsaturated fatty acids, and other uniquely eucaryotic substances require molecular oxygen and these biochemical capabilities evolved with the changing environment.

According to Dayhoff and her associates[5] the eucaryotic branch emanated from the middle portion of the phylogenetic tree, with two bacterial species, *Halobacterium salinarium* and *Thermoplasma acidophilum,* diverging from the host line. This is supported by the similarity of the protein-synthesizing system of *Halobacterium* to that of eucaryotes, and the fact that *Thermoplasma* lacks a cell wall, has histone-like proteins to protect its DNA, contains actin-like protein, and uses flavin oxidases for its oxygen metabolism. This interpretation, however, is not accepted by all biologists.[6] Carl Woese[7] at the University of Illinois believes that these unusual species are much older and represent a third kind of life that thrived when the earth was far different from what it is today.

Woese also used molecular sequencing to work out the phylogeny of bacteria but has based his findings exclusively on the comparison of a specific ribosomal RNA. Ribosomes consist of RNA molecules complexed with protein, and typically a bacterial cell has between 10,000 and 20,000 ribosomes. Because ribosomal RNAs have remained constant in function throughout the existence of life, some portions of the molecules have changed so slowly over the eons that it is unlikely that the common ancestral sequence has been totally obliterated. Exact copies no longer exist because mutations have altered the original sequence, but there should remain in these macromolecules vestiges of the original ribosomal RNA and its gene that have survived evolutionary time.

There are three ribosomal RNAs, designated 23S, 16S, and 5S, based on their rate of sedimentation in the ultracentrifuge.* Their respective number of nucleotides is 2,900, 1,540, and 120. For characterization, the small 5S rRNA is not as accurate an indicator of phylogenetic relations as the longer RNAs. And because the 23S rRNA sometimes exhibits large, anomalous differences in sequence from one species to another, the 16S ribosomal RNA has been the molecule of choice for characterization.

When Woese began to explore bacterial relations in 1969 by comparing the

---

* S stands for Svedberg, a unit of measure for the rate of sedimentation in an ultracentrifuge and an indirect measure of molecular weight.

sequences of 16S ribosomal RNAs, the technology to determine the nucleotide sequence of the entire molecule was not technically feasible, as it is now. It was possible, however, to sequence short segments, or oligonucleotides. The technique involved labeling the 16S ribosomal RNA from a given species with radioactive phosphorus-32 and using the enzyme ribonuclease to digest the molecule. The digest was then fractionated into the oligonucleotides by two-dimensional paper electrophoresis and each was sequenced, producing a catalog of sequences for the organism. Comparing catalogs from various organisms was like comparing their 16S rRNA sequences and could be used to work out the genealogical relations.[8]

Digestion of 16S ribosomal RNA with T_1 ribonuclease cleaves the molecule into a number of segments with some 15–20 residues long. Oligonucleotides six units or more in length account for 500–600 residues, or 35 percent, of the 1,540 nucleotides in the molecule. Those oligomers containing more than five nucleotides almost always occurred only once as a sequence in a 16S rRNA molecule. When, therefore, 16S rRNAs from different organisms contained the same six-unit sequence, the probability of it being a true homology was almost certain.

The ribosomal RNAs of almost 200 species of bacteria and eucaryotes have been characterized by the University of Illinois group. In the course of screening bacteria in collaboration with Ralph S. Wolfe of the microbiology department an unforeseen relationship emerged. The methanogens—organisms that live in oxygen-free environments and generate methane by the reduction of carbon dioxide—did not fall into the phylogenetic group with the other bacteria. This was an unexpected result. Methanogens look like bacteria and have always been regarded as members of that group: they are the same size, they have no nuclear membrane, and their DNA is small like that of bacteria. Yet the comparison of 16S ribosomal RNAs indicated that methanogens are no more closely related to the bacteria than to the eucaryotes.

Further analysis gave the same result for two other types of organisms: the extreme halophiles and the thermoacidophiles. The extreme halophiles are bacteria that require high concentrations of salt and thrive readily in saturated brine. They are found along ocean borders and in the Great Salt Lake and the Dead Sea. They also give a red color to salt evaporation ponds and are responsible when salted fish spoil.

The thermoacidophiles, on the other hand, are notable for their tolerance for high temperatures and strongly acid conditions. One genus—the *Sulfolobus*—is found in hot sulfur springs where the temperature can be above 90°C and the pH below 2. And another genus, the *Thermoplasma,* has been found in the smoldering piles of coal tailings.

Phylogenetic differences between these organisms and the typical bacteria should be reflected in biochemical differences, and this has been found to be the case. A general feature of bacteria is their cell wall. This cell wall contains the sugar derivative muramic acid, which is a basic unit of a complex polymer called peptidoglycan. It was known that the extreme halophiles and the thermoacidophiles were exceptions to this generalization, but it was assumed that it was merely an idiosyncracy. When a systematic comparison was made, however, it was discovered that none of the methanogens, extreme halophiles, or thermoacidophiles had the typical bacterial murein-type cell wall. Instead, within the group there exists a number of quite different wall types. Some have pseudomurein walls resembling typical bacterial walls but not containing muramic acid,[9] some have walls made of protein subunits,[10] some of polysaccharides,[11] and some, the mycoplasma, have no walls at all.[12]

When the cellular membranes of the group were analyzed, it was found that their lipids were distinctly different. The lipids of both bacteria and eucaryotes consist of straight chain fatty acids joined to a glycerol molecule through an ester linkage (-CO-O-). The lipids of the methanogens, the extreme halophiles, and the thermoacidophiles are also glycerides, but they differ by having two long hydrocarbon chains joined to glycerol by an ether linkage (-O-). In other words, instead of being a diester of fatty acids, the lipids are diethers composed of glycerol and two molecules of the alcohol phytanol.[13]

This small group of eccentric organisms differs from other forms of life by yet another clear distinction. Basic to the synthesis of proteins in all organisms is a series of transfer RNA molecules that carry the particular amino acids to the messenger RNA for attachment to the peptide chain. In all organisms the transfer RNAs have a uracil that has been methylated to form thymine—except in methanogens, extreme halophiles, and thermoacidophiles. In these organisms the uracil has not been converted to thymine but rather has been modified to form either pseudouridine or another yet to be identified nucleotide.[14]

These findings have led Woese to propose that there are not two kinds of life, but actually three. He calls the third group archaebacteria.[15] If there are then three primary kingdoms—the archaebacteria, the true bacteria, and the eucaryotes—this raises a new question. At what stage in evolution did the different forms of life diverge, and what was the common ancestor?

In terms of the ribosomal catalogs, the archaebacteria, the true bacteria, and the eucaryotes appear to be equidistant from each other in evolutionary time. Instead of eucaryotes arising from a union of procaryotes, it seems possible that archaebacteria may also have participated in the endosymbiotic creation. At least one of the eucaryotic cytoplasmic genes is different from the others. The eucaryotic nucleus apparently contains three kinds of genes, which presumably are from the genetic material of the host cell and from genes expropriated from the genomes of the organelles. The gene for the ribosomal A protein, however, has been found to be not of procaryotic derivation but of archaebacterial origin.[16]

When the archaebacteria are excluded, despite their apparent diversity, the bacteria as a class have a close phylogenetic relationship. There are eight major bacterial divisions that appear to have diverged from one another over a relatively short period of time so that their exact order is difficult to establish. They

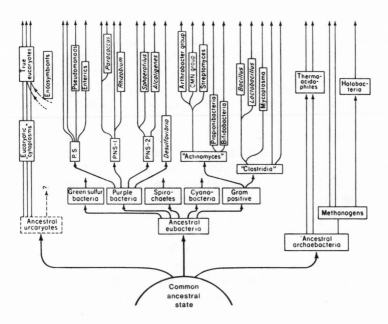

Figure 12.2. Schematic representation of the major lines of procaryotic descent.

nevertheless have the same type of cell wall, and their protein-synthesizing mechanism is quite standard for both the procaryotes and the eucaryotes.

The archaebacteria, on the other hand, despite the small number of known species, have a diversity as measured by phylogenetic depth and phenotypes to be equal to or greater than that found among the bacteria. There are at least four major cell wall types, in contrast to the one for bacteria. And whereas the subunit structure of the RNA polymerase is constant among the bacteria, the polymerase structure for the archaebacteria is different.[17] This may mean that the time when bacteria and archaebacteria separated from a common ancestor is so ancient that even the function of RNA polymerase was still in the process of being refined.

The archaebacteria are notable for the extreme environmental niches that they occupy. The methanogens are widely distributed but are not commonly encountered because they are killed by oxygen. They live only where carbon dioxide and hydrogen are available, which means they exist in close association with other bacteria that produce these gases, such as clostridia that metabolize decaying organic matter and give off hydrogen as a waste product. Methanogens, therefore, are found in sediments of bogs, streams, and lakes rich in decaying vegetation, where they are responsible for "marsh gas." They also are found in the intestinal tracts of animals in general and in the rumen of cattle and other ruminants. And they can be isolated from ocean bottoms and hot springs.

The extreme halophiles live under conditions too drastic for most other organisms. They thrive in waters of high salt concentrations, where they have to maintain large concentration gradients across their cell membrane to keep their internal ion level from being excessive and to move substances in and out of the cell. Halophiles have a relatively simple photosynthesis mechanism that is not based on chlorophyll but on a membrane-bound pigment called bacterial rhodopsin.

The thermoacidophiles also have to maintain a sizeable gradient across their cell membrane, but it is a pH gradient to sustain a near neutral pH inside the cell while living in strong acid solutions. This gradient is dependent on the high temperatures of sulfur springs where they live, and when *Sulfolobus* is placed in a lower temperature its metabolism stops. No longer able to maintain its gradient for internal neutrality, the organism dies.

The archaebacteria belong to a class of organisms that thrive under conditions that seem harsh and forbidding to other forms of life. It has generally been assumed that they represent the expansion of the living world into the fringes of the life zone through adaptation. The answer, however, may be that they are refugees from a changing world where they once held dominion. Woese contends that the unusual metabolism and control mechanisms of the archaebacteria probably reflect the conditions that once were prevailing when life was first appearing on a youthful earth.[18]

13
Energetics

We all have an ancestry that extends in an uninterrupted line back more than 3.4 billion years. It was at some moment in that incredibly distant past that substances on primordial earth came together to form an entity that was able to absorb energy and use it to grow and assemble ever more complex units. Unlike the inanimate world that absorbs energy only to become more disordered, the biological systems that formed had a capacity to direct energy toward greater organization and expansion. This capability that we inherited from the first living cells characterizes all living things.

A biological cell performs the synthesis of its organic constituents by a series of biochemical reactions catalyzed by enzymes. How an enzyme possessed such impressive catalytic powers remained a mystery for decades; but there persisted a more fundamental problem. A biological system synthesizes its components by building complex molecules from simple chemicals. They grow and multiply, reproducing not one or two types of biochemicals but literally thousands, all meshed in a highly organized manner. These are all functions that require energy. In effect, life creates order from disorder.

The basic concepts of energy were formalized in the 1840s, long before the life processes were understood. The first law of thermodynamics recognizes the conservation of energy, the principle that neither matter nor energy can be gained or lost within a closed system. The second law describes the flow of free energy, ΔG, within a system and the universe, as well as order, that is, entropy. Energy flows from areas of higher concentrations to those of lower. Like water, energy seeks a level, as witnessed by the flow of heat from hot objects to cold.

This direction of energy flow applies to all conditions of internal energy. The order created in constructing skyscrapers or complex molecules from a less organized distribution of matter represents an investment of energy. But concentrations of energy in organization standing above the general level cannot be retained indefinitely without an input of energy to resist the tide toward equilibration.

Any process concentrating matter to build order, therefore, is going against the energy tide. Nevertheless, this is exactly what biological systems do. To see how it is possible, we may regard the biological system as a localized phenomenon, best understood by comparing its behavior with the analogous behavior of water. Despite the fact that water tends to seek a level, there are waves that rise above the level to an unstable position at the expense of the downward trough. So too does the randomness of the universe tend to seek a common level of thermodynamic equilibrium, but waves of order rise above the equilibrium level at the expense of the energy of other substances.

The overall procedure for this does not controvert the second law. Plants, the initial collectors of solar energy, absorb more energy from the sun than they store as chemical energy, so that the total process is a downhill chain of events. Animals then consume plants to build their components, and it is the stored chemical energy of plants that supplies the metabolic energy for animals.

Because of energy barriers restricting the flow of chemical reactions, there exists a narrow temperature range where durable molecular configurations can be rather complex. Molecules can form under short-lived conditions of high energy and freeze into chemical structures with potential energies. Some of these molecules can interact chemically to absorb energy and direct it to elevate the internal energy of elaborate cellular structure and components. This is the principle of physics upon which life rose, using the energy of the sun to maintain itself.

All biological reactions are spontaneous reactions. This means that the reactions are uphill from equilibrium and energetically favored to move toward equilibration, although it tells nothing about the rate of the reactions. Like a rock on a hillside, biological reactions tend to be static unless given a push. Gasoline or sugar, for example, can stand in the air almost indefinitely with no perceptible signs of combustion, despite the fact that their oxidation is a vigorous exothermic reaction. For the reaction to proceed, the weaker C-H bonds need to break before the stronger C-O and O-H bonds can form. At room temperature, these C-H bonds are quite stable. If the reaction is to go quickly, sufficient energy, called the activation energy, must be added to break the C-H bonds. Once this is accomplished, the energy from the reaction is sufficient to fuel the process in a chain reaction.

Spontaneous reactions have the potential energy to continue on their own, but do so only if the activation energy barrier is overcome. Enzymes lower the barrier by forming a transient intermediate complex with the reactant that weakens the bond that forms the barrier. This capability not only makes enzymes highly specific catalysts, it makes them more effective than acid or base catalysis. Because of the activation energies of the chemical reactions involved in cellular activities, the life processes can be rigidly controlled by enzymes. If spontaneous reactions did not have energy barriers but proceeded instantaneously, life would not have been possible.

Despite all of this, the principal activity of biological systems and the basis of growth and reproduction is synthesizing complex molecules from small compounds. These syntheses are not spontaneous reactions, but require an input of energy. They are on the downhill side of equilibrium. The products have more

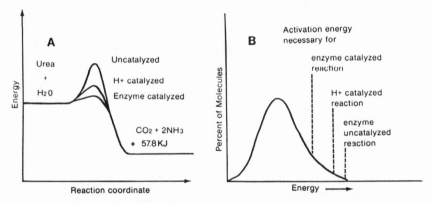

A. Energy diagram showing the effect of catalysis on the activation energy in the hydrolysis of urea. B. Diagram showing the effect of lowering the activation energy on the percent of molecules having sufficient energy for reaction.

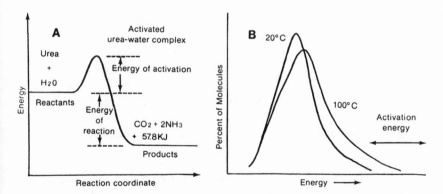

A. Energy diagram of the hydrolysis of urea. Although the formation of products liberates 57.8 KJ/mole, the reaction cannot take place without overcoming an activation energy. B. Distribution curves of activation energy at 20° and 100°.

Figure 13.1. Energy diagrams of the decomposition of urea.

energy than the reactants. These, therefore, are not reactions that go spontaneously and consequently cannot be promoted by a catalyst. But to a casual observer it appears that a biological system does just that.

How then is it possible?

The manner in which a cell carries out chemical conversions began to become evident between 1930 and 1940, when the chemical details to the biochemical degradation of glucose were discovered. Biochemists realized that there had to be some method in which the cell accepted chemical energy and transferred it from one compound to another. The biochemical agents that play this role were found to be the phosphate compounds adenosine triphosphate (ATP) and adenosine diphosphate (ADP).

In 1941, Fritz Lipmann[1] introduced the concept of the high-energy phosphate bond, and the role of these compounds in bioenergetics began to be clearly appreciated. ATP is a compound of high potential energy. When it loses a phosphate group by hydrolysis to form ADP and inorganic phosphate, a rather large amount of energy is liberated. Under standard conditions of pH 7 and 25°C, the standard free energy change, ΔG, for the hydrolysis of ATP is -30.5 kilojoules.*

$$ATP + H_2O \longrightarrow ADP + P_i \quad \Delta G = -30.5 \text{ KJ}$$

The free energy change of ΔG is the energy that drives the reaction, while giving off heat to the surroundings. The value of energy change is minus when heat is liberated because it represents a reduction in the overall energy content. For a system in equilibrium at constant temperature and pressure, $\Delta G = O$. A spontaneous reaction will have a negative free energy change, and the higher the negative value, the greater is the driving force of the reaction.

If, then, the cell wants to carry out a synthesis in which the product has an energy level higher than that of the reactants, that is, an uphill reaction, the reaction is changed to a spontaneous reaction by making an activated derivative of the reactant. This is achieved by transferring a phosphate to a reactant through phosphorylation by ATP. An example is the biosynthesis of glutamine, an important amino acid in proteins. To form the amide bond with glutamic acid and ammonia requires an input of 14.2 kilojoules of energy. On the other hand, glutamic acid can be converted to the mixed anhydride by reacting with ATP in a spontaneous reaction. This derivative, glutamyl phosphate, now is of a higher energy level than glutamine, so the reaction with ammonia to form glutamine becomes a downhill reaction that goes spontaneously. The combined effect of converting glutamic acid to glutamine of a higher energy through the use of ATP of an even higher level gives a net release of 16.3 kilojoules of energy, enough to drive the reaction well toward the product.

---

* Actually, under cellular conditions, Mg^{++} forms complexes of differing affinities with ATP, ADP, and phosphate. This, and other factors that affect the equilibrium of the reaction, tend to increase the energy of hydrolysis of ATP to an even higher value.

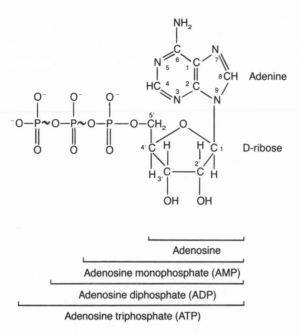

Figure 13.2. The structure of adenosine phosphate derivatives.

This is the basic procedure the cell uses to synthesize ester linkages, peptide bonds, and glycosidic couplings—all endothermic reactions that require energy. It is the chemical energy of the phosphate bond of ATP, far in excess of energy added to the products, that is the source of energy for these syntheses.

To carry out its many syntheses, a cell needs to continually generate ATP. Plants with their photosynthetic mechanism produce their ATP through a complex procedure using the energy of absorbed sunlight, which, in turn, they utilize to reduce carbon dioxide to carbohydrates as stored chemical energy. Plants also have mitochondria that permit them to generate ATP through the breakdown of their carbohydrate reserves.

The amount of energy in a mole* of glucose is 2,870 kilojoules, the quantity of heat liberated when glucose is oxidized by oxygen to carbon dioxide and water.

$$C_6H_{12}O_6 + 6\ O_2 \longrightarrow 6\ CO_2 + 6\ H_2O \quad \Delta G = -2,870\ KJ$$

Animals, using plants as food, draw from this potential energy stored in the chemical bonds of sugar to generate their ATP and to use the heat to maintain their body temperature. To do so, they do not want all the energy liberated as

* One mole is the numerical value of the molecular weight expressed in grams. A mole of glucose is 180 grams.

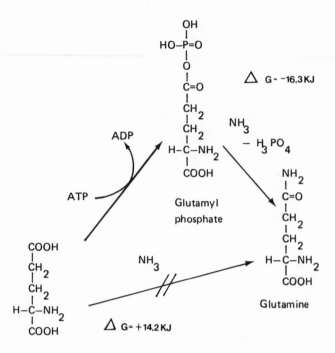

Figure 13.3. The biosynthesis of glutamine.

Table 13.1. Standard free energy of hydrolysis of phosphate compounds.

	ΔG (KJ/mole)	Direction of phosphate transfer
Phosphoenolpyruvate	-61.9	
1,3-Diphosphoglycerate	-49.4	
Phosphocreatine	-43.1	
Acetyl phosphate	-42.3	
ATP	-30.5	
Glucose 1-phosphate	-20.9	
Fructose 6-phosphate	-15.9	
Glucose 6-phosphate	-13.8	
3-Phosphoglycerate	-10.0	
Glycerol 3-phosphate	-9.2	

heat, but instead, as much as possible to be collected in high energy chemical bonds of compounds like ATP which they can use in their syntheses.

One reason ATP was adopted by living systems as the agent for the transfer of the phosphate group is not that it is a phosphate derivative of the highest potential energy but because it is of an intermediate level among phosphate compounds. To function as a phosphate transfer agent, ATP must be able to accept phosphate groups from some compounds and donate them to others. There then exists in organisms a group of phosphate compounds comprising a scale of free energy. In biological processes, when ATP is generated it receives its phosphate from a phosphate derivative more energetic than itself in order for the reaction to be spontaneous.

14
The Driving Force

A biological system is more than an assembly of organic substances. It is a mechanism that can take up energy and use it to elevate simple chemicals to more complex arrangements. The success then in life's creation was the coming together of prebiotic materials that could act in concert to harness available energy in the environment. All the energy—ultimately derived from the sun—was present in two forms. It existed as chemical energy sealed in the structure of organic substances produced by the various forms of energy acting on the atmosphere, and it was as the sunlight that bathed the earth's surface each day. Eventually, both sources of energy were used by organisms.

Of all the biological processes studied by chemists, few have elicited more wonder or given greater resistance to being unraveled than photosynthesis. With fermentation, all the enzymes and other components in the breakdown of glucose to pyruvic acid have been isolated, and each step, as well as the complete series of reactions, has been reproduced in the laboratory. With photosynthesis, the case is quite different. When extracts are made of plant cells to collect chlorophyll and the other components, the extracts are no longer capable of photosynthesis. In the view of biologists, this is like smashing a watch and looking at the pieces to see how it works, rather than observing the complete mechanism.

But considerable research has been devoted to studying photosynthesis, much of which is now well understood. Since light is a form of energy, plants have, in effect, developed a biochemical means of trapping the energy of sunlight and converting it directly to chemical energy. They have succeeded in channeling it to the synthesis of ATP and the reduction of carbon dioxide for the production of carbohydrates.

To reduce carbon dioxide to the reduction level of sugar takes an input of $+468$ KJ/mole of carbon dioxide.

$$CO_2 + 4H \longrightarrow CH_2O + H_2O \quad \Delta G = +468KJ$$

Light travels in waves of particles known as photons. Rather than being continuous, the energy is in discrete packets called quanta, the energy of each quantum varying inversely with the length of the photon wave (λ). That is to say, as the wavelengths become longer, going from ultraviolet to visible to infrared, the energy of a quantum of light becomes less. In the mathematical expression of a light quantum, ν is the frequency, h is Planck's constant, and c is the velocity of light.

$$E \text{ (light quantum)} = h\nu = hc/\lambda \quad \text{thus } E \propto 1/\lambda$$

Red light with its wavelength of 0.68 micrometer can stimulate green plant photosynthesis by a light quantum corresponding to 2.9×10^{-12} ergs. When this quantity of energy is converted to units of chemical reactions, the energy compares to 167 KJ/mole. This means it takes at least three quanta of light to reduce a mole of carbon dioxide. In actuality it seems to take ten.

Generally, when light is absorbed by dark or colored objects, they become warm as energy is converted to heat. But just as biological systems have developed means of tapping the energy from the oxidation of organic substances, so too have they succeeded in draining off part of absorbed light energy before it is lost as heat.

If the photons are of short wavelength, such as for ultraviolet light, and consequently of high energy, they are destructive to many organic compounds by cleaving covalent bonds. On the other hand, if the wavelengths are too long and of low energy, they are too low a frequency to be absorbed. In order for energy in the form of a wave such as light to be transferred, it can pass only to something that is vibrating at the same frequency. Electrons, like all matter in motion, travel in a wave pattern of a definite frequency that depends on the mass and velocity. The frequencies of most electrons around atoms are extremely short, of the order of X-rays, but in molecules with alternating single and double bonds, the electrons that resonate the length of a series can be of long wavelengths. These compounds are the colored organic substances that absorb the blue, red, or yellow light and reflect the complementary color.

The orbits of electrons are not fixed but are divided into subunits or orbitals that represent a range of energy levels within an orbit. When, therefore, a molecule absorbs a light quantum, the energy is passed to an electron which is elevated to an orbital of a higher energy level. At this point, it is said to be in the excited state. Excited states, however, are unstable configurations and the electron quickly drops back to the lower level, or ground state, dispensing with the energy by emission of light (fluorescence) or through collisions, generating heat.

In order to harness energy to do work, it must not escape as heat or light, but

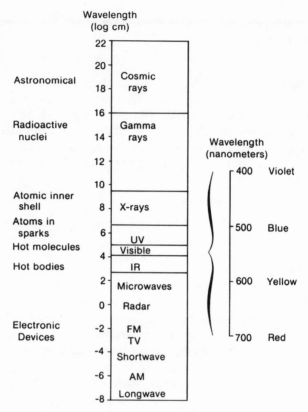

Figure 14.1. The electromagnetic spectrum.

must be funneled to a usable energy form, which for biological systems means chemical energy. When electrons are energized sufficiently, they can often be induced to leave the orbit of the excited molecule and transfer to an acceptor substance. This then becomes the oxidation-reduction chemical reaction where an oxidant accepts electrons from a donor and is reduced by it. It is the mainspring of photosynthesis and the manner in which sunlight is converted directly to chemical energy.

Plants contain a number of pigments such as carotenes and carotenoids (which give the yellow-orange color of carrots) that absorb light, although the chlorophylls are the main light-absorbing molecules in green plants. When chlorophyll absorbs a quantum of sunlight, the energy of an energized chlorophyll will be quickly reemitted through fluorescence unless the electron is transferred to a more stable state.

In 1937, Robert Hill[1] of Cambridge University made a discovery that led to the identification of a number of electron carriers that function to transport electrons away from excited chlorophyll. Hill found that dried or powdered leaves,

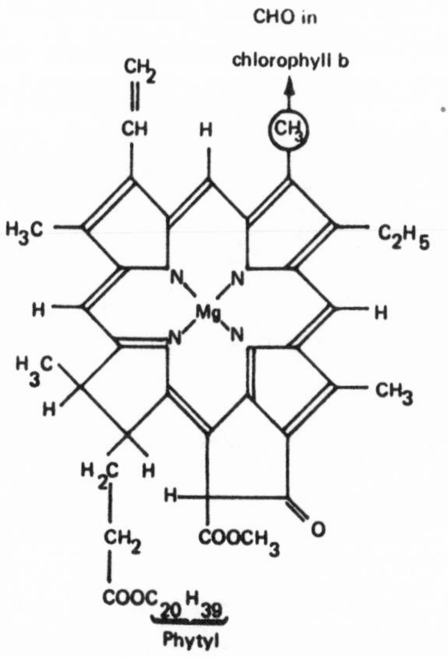

CHO in

chlorophyll b

CH_2

CH H CH_3

H_3C C_2H_5

N N

H Mg H

N N

H_3C

H CH_3

H_2C H

H

CH_2 $COOCH_3$ O

$COOC_{20}H_{39}$

Phytyl

Figure 14.2. The structure of chlorophyll a. The phytyl group is a long, unbranched chain.

suspended in water with ferric oxalate or some other ferric salt and illuminated, released oxygen for an hour or longer—much longer than without a ferric salt. Later studies showed that the ferric salts could be replaced by quinone or certain dyes. In all cases the added substances had one thing in common—they were strong oxidants. Ferric salts were acceptors of electrons, and the particular organic dyes were acceptors of hydrogen atoms.

$$Fe^{+++} + e^- \longrightarrow Fe^{++}$$

This observation sparked a search for naturally occurring electron carriers in plants that may be connected with photosynthesis. Ultimately, two important electron acceptors were found. They were coenzymes and chemically related to the building blocks of the nucleic acids. These were nicotinamide adenine dinucleotide (NAD) and its phosphorylated derivative (NADP).

It is the nicotinamide portion of the molecule that can accept and give up electrons. Just as ATP is a phosphate carrier, so too are NAD and NADP carriers of electrons. In man and other vertebrates, the nicotinamide part of the compound must be supplied through the diet as niacin, one of the B vitamins.

Although similar in chemical composition, NAD and NADP each play a specific role different from the other. NAD accepts electrons that are directed

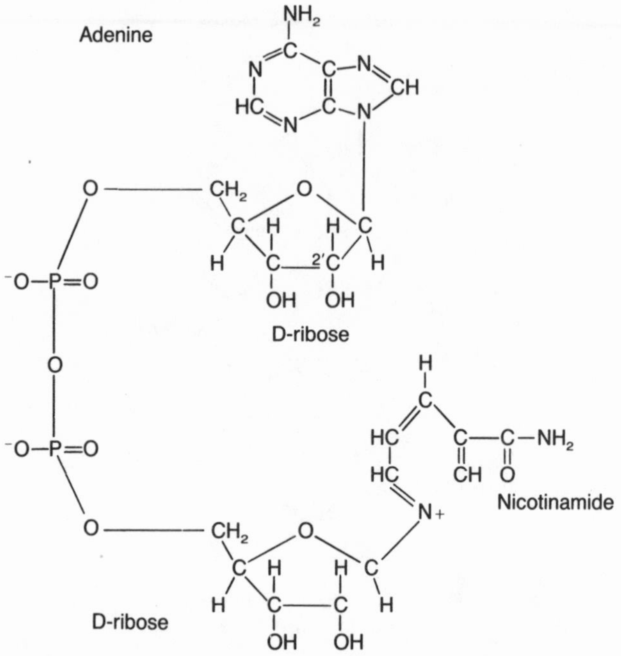

Figure 14.3. Nicotinamide adenine dinucleotide (NAD). NADP has a phosphate group attached to the 2-hydroxyl of the ribose next to the adenine ring.

toward oxygen in respiration, whereas NADP directs its electrons for the reduction of organic compounds. NADP, therefore, is more immediate in the reduction of carbon dioxide to sugar.

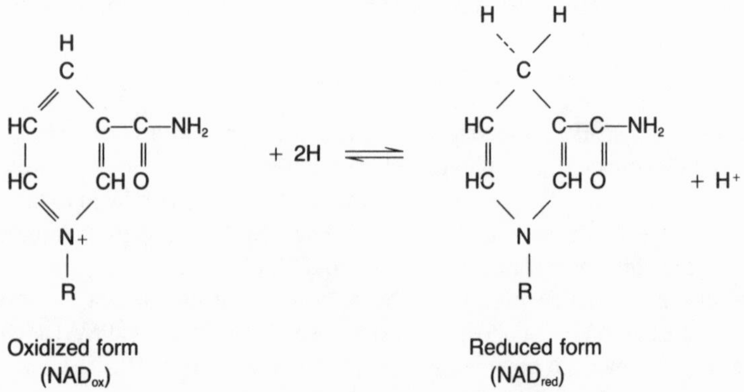

NADP does not receive electrons from energized chlorophyll directly, but instead, the electron is transported by a series of specialized compounds. It is at this point that ferredoxin, the iron-sulfide protein that may have been the first protein used by living cells, becomes important. When an electron is energized in a chlorophyll molecule it leaves this compound and passes to another type of chlorophyll molecule designated P700. The energized electron is pulled away from P700 by Z, a strong electron acceptor. Z then passes the electron to ferredoxin, whose iron is reduced from Fe^{+++} to Fe^{++} in the process. From ferredoxin the electron is passed by an electron carrier—ferredoxin-NADP oxido-reductase—to $NADP_{ox}$, converting it to $NADP_{red}$. As ferredoxin gives up the electron, its iron is oxidized from Fe^{++} back to Fe^{+++}. Thus there is a light-induced flow of electrons from chlorophyll to NADP until all the available NADP is reduced.

Electron transport chain

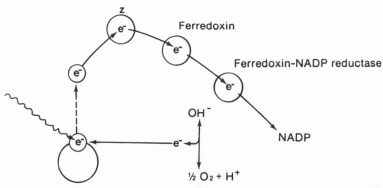

Figure 14.4. The high-energy electrons from energized chlorophyll are used to reduce $NADP_{ox}$ to $NADP_{red}$.

Once NAD or NADP is reduced, this completes the link from the absorption of sunlight to enzymic synthesis, for these reduced coenzymes can then participate in the various biochemical conversions, including carboxylations, which are dependent upon reduced pyridine nucleotides.

It was generally assumed that photosynthetic organisms obtained all their ATP from the breakdown of glucose, an implication that glycolysis or similar degradations would have preceded the biosynthesis of ATP in the origin of life. But in 1954, Daniel Arnon with Mary Belle Allen and Frederick Whatley[2] at Berkeley discovered what has come to be called photophosphorylation. They observed that chloroplasts, unaided by other cellular particles, have the ability for carbon dioxide fixation and for the direct conversion of light energy into ATP.

This is the simplest—possibly the original—harnessing of the sun's radiant energy to produce ATP. The ATP formation was in complete absence of organic substrates or oxygen. When isolated chloroplasts were illuminated in the presence of ADP and inorganic phosphate, ATP formed at a high rate. The longer the chloroplasts were illuminated, the greater the amount of ATP formed.

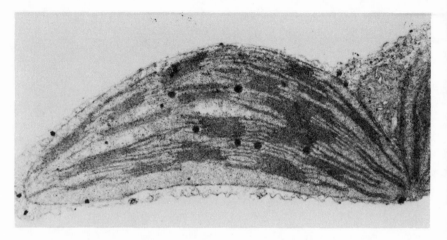

Figure 14.5. A tomato chloroplast as viewed through the electron microscope. The striated sections are the lamellar vesicles with membranes that serve as the essential insulating barriers for photosynthesis. Magnification is 32,000 X.

In photophosphorylation, an external electron donor is not needed. The generation of ATP is due to a cyclic flow of electrons from the excited chlorophyll to ferredoxin, then through an enzymic mechanism, and back again to chlorophyll by a chain of carriers. In order for this process to take place, the first electron donor has to be separated or insulated from the last electron acceptor; otherwise, the circuit of the electron flow would be shorted. This separation is created by the insulating barrier of the lipid membrane. This is why photosynthesis, unlike an extract for anaerobic glycolysis, cannot occur in isolated solutions. The structural integrity is an absolute requirement for photophosphorylation, and life would not have been possible without the unique properties of the lipid membrane.

The photosynthetic organisms are able to generate ATP by capturing and using the energy of sunlight directly. This may well have been the initial manner of the primitive cells of producing their ATP. The degradation of glucose, however, is also of early origin and a model of simplicity. In photosynthesis, organisms convert solar energy to chemical energy by building organic compounds of a higher energy level. In glycolysis, organisms take glucose containing that chemical energy and degrade it to derivatives of lower energy, using the energy difference to produce ATP and to drive the reaction steps.

When Büchner in 1897 found that chemical components from yeast were able to break down glucose to ethanol and carbon dioxide outside the cell structure, he discovered the original procedure of biological systems for extracting chemical energy from organic compounds. At the time, Büchner believed fermentation was a single chemical reaction catalyzed by an enzyme he called zymase, but when the mechanism of the conversion was eventually solved in its entirety, it was learned that the breakdown of glucose consisted of eleven separate reactions, each catalyzed by a specific enzyme. This is the Embden-Meyerhof pathway, named after the German biochemists who postulated and worked out the critical steps of the series in the late 1920s and early 1930s.

Brewer's yeast ferments glucose to ethanol and carbon dioxide, but different bacteria generate acetone, butanol, acetic acid, and ethanol as their end products of fermentation. The principal breakdown, however, is the anaerobic degradation of glucose to pyruvic acid and lactic acid. This is the initial degradation of glucose that exists in all living things, inherited from the primitive anaerobes that were the first forms of life on earth. The reaction sequence has been extended during evolution by different organisms to give various products. In animals, the anaerobic pathway has survived in muscle contraction where pyruvic and lactic acids appear as primary end products before oxygen is admitted for further oxidation to carbon dioxide and water.

$$C_6H_{12}O_6 \xrightarrow{\text{11 steps}} 2 \begin{array}{c} COOH \\ HO-C-H \\ CH_3 \end{array} \quad \Delta G = -197 \text{ KJ}$$

Glucose   Lactic acid

As Büchner discovered, the entire process of fermentation proceeds merely by having all the reactants together in solution. Since the conversion has been found to be a series of separate reactions, each reaction capable of being conducted independently in the laboratory, the products of each step serve as the reactants, or substrate, for the enzyme of the succeeding reaction. With all the components present, the breakdown of glucose, therefore, proceeds nonstop through each step to the final product.

To carry out glycolysis, glucose is phosphorylated with ATP, so that in the breakdown, two derivatives are produced that are of a higher energy level than ATP. Consequently, these high-energy derivatives are able to generate ATP by passing their phosphate group to ADP. The overall result in the degradation of glucose to lactic acid is that two ATP molecules used to phosphorylate glucose lead to the formation of four ATPs from ADP for a net yield of two ATPs. In the process, NAD_{ox} is reduced to NAD_{red}, but is regenerated in the end when it converts pyruvic acid to lactic acid. The combined equation of all the reactions becomes

Glucose + 2 P_i + 2 ADP $\longrightarrow$ 2 Lactic acids + 2 ATP + 2 H_2O

When glucose is broken down stepwise to yield chemical energy for the production of ATP, there is one chemical reaction that is the key to the energy transfer. It is the oxidation of glyceraldehyde 3-phosphate by NAD_{ox} with the simultaneous phosphorylation, elevating the intermediate to 1,3-diphosphoglycerate, a high-energy phosphate derivative whose energy is higher than ATP. In the final reaction of the pathway, NAD_{ox} is regenerated in the reduction of pyruvic acid to lactic acid by NAD_{red}, thus completing the cycle of the oxidation–reduction reaction. It was at this step of oxidation by NAD_{ox} that some of the chemical energy invested in glucose by the reduction of carbon dioxide through phosphorylation was recovered by the reverse reaction of oxidation.

What then is the basis of oxidation–reduction that permits it to play a central role in bioenergetics?

Except for fluorine, oxygen is the most electronegative element, whereas hydrogen is electropositive. This electrochemical opposition of the two elements makes a reaction between oxygen and hydrogen one of the most vigorous, and the O-H bond one of the strongest. An oxidation reaction, typically characterized by the addition of oxygen, is simultaneously accompanied by the reduction of the oxygen. With the formation of water, the reductant is hydrogen. In a broader sense, the reductant is anything that is being oxidized. Since a hydrogen atom consists simply of an electron and a proton (H^+), the addition of an electron is a reduction, its removal, an oxidation.

Through photosynthesis plants use the energy of sunlight to reduce carbon dioxide to glucose with an input of 2,870 KJ/mole of glucose. It then is this reservoir of chemical energy that organisms draw upon to produce ATP. To liberate 2,870 kilojoules, the glucose would have to be oxidized completely with oxygen back to carbon dioxide and water. This procedure, however, was inaccessible to the Archean procaryotes when little or no free oxygen existed in their environment.

Nonetheless, some of the energy can be withdrawn because atoms can exist in various states of oxidation, depending upon the composition of the compound. For carbon, carbon dioxide (CO_2) is the fully oxidized state, and methane (CH_4) is the fully reduced. All other carbon compounds represent intermediate levels of oxidation. By converting one compound to another where the carbon is less reduced represents an oxidation, and the difference in free energy between the compounds is liberated. Accordingly, when glucose is broken down to lactic acid by the anaerobic pathway, the oxidation liberates 197 kilojoules of the 2.870 kilojoules in each mole of glucose. In the process, two molecules of ATP are produced from one of glucose, each with an energy of 30.5 kilojoules. The pathway, therefore, captures a total of 61 kilojoules of the 197 kilojoules of energy, using the remaining 135.6 kilojoules as the driving force to assure completion of the reactions. Of the total of 2,870 kilojoules in the structure of glucose, the anaerobic organisms were able to recover only 61, or about 2 percent.

The anaerobic breakdown of glucose to pyruvic acid provided an extractable source of chemical energy for producing ATP. The supply of abiotic carbohydrates eventually proved exhaustible, and organisms developed a series of reactions to reverse the process to synthesize glucose from pyruvic acid. Seven of the steps in glycolysis are reversible; the remaining three reactions are bypassed by enzyme-catalyzed conversions that are made spontaneous in the direction of synthesis by an input of six high-energy phosphate bonds, four from ATP and two from GTP (guanosine triphosphate).

Just as biological systems developed means of using ATP to rebuild their stocks of glucose, so also did they use carbon dioxide as the source of carbon to synthesize glucose. Early organisms were able to derive ATP by photophosphorylation where they harnessed the energy of sunlight, and using ATP as an activator, they were able to degrade glucose and other organic compounds to tap their potential chemical energy. Glucose was a store of chemical energy and could be rebuilt from pyruvic acid with an input of ATP. But for continuing their growth, biological systems needed to draw from a supply of carbon and nitrogen when the available organic compounds became exhausted. They found a virtually inexhaustible reservoir in the atmosphere.

The carbon dioxide in the atmosphere was readily accessible, but chemically it was fully oxidized and needed to be partially reduced before it could be assimilated as organic material. In order to do this, photosynthetic organisms developed a means of "CO_2 fixation" by using the reducing power of photosynthesis. They had $NADP_{red}$ for a reductant and ATP for supplying the driving force; a pathway of spontaneous reactions that was needed was evolved to incorporate carbon dioxide into organic compounds.

$$6\ NADP_{red} + 12\ H_2O + 12\ ATP + 6\ CO_2 \longrightarrow$$
$$C_6H_{12}O_6 + 6\ NADP_{ox} + 12\ ADP + 12\ P_i$$

The reducing power of photosynthesis that is used to convert carbon dioxide to glucose produces at the same time an extremely strong oxidizing potential (oxidant) that extracts hydrogen from an external donor. The earliest forms of life, with a photosynthetic mechanism less evolved than the oxygen-liberating blue-green algae, would have extracted their hydrogen from available sources that required the least amount of energy to give up their hydrogens. The external source of electrons may be organic or inorganic. When isopropanol is the donor, it is subsequently oxidized to acetone; similarly, succinate becomes fumarate. Inorganic electron donors used by some bacteria are the sulfides.

Ultimately, the substance never used by bacteria which became the electron source for blue-green algae and all plant life is water.

$$2\ H_2O \longrightarrow 4\ H^+ + O_2 + 2\ e^-$$

It takes ten times as much energy to extract hydrogen from water than from hydrogen sulfide, but either because of a dwindling supply of the early donors or perhaps due to the sheer abundance of water, photosynthetic organisms evolved that were capable of wrestling hydrogen from oxygen in water molecules. This came to be a momentous event in the evolution of life on earth—for the result was the discharge of oxygen as a by-product into the environment.

It was the release of oxygen into the atmosphere that eventually led to the development of the third means of producing ATP. All biological systems were able to degrade glucose to pyruvic acid by the anaerobic pathway to extract some of the chemical energy, but most of the energy remained in the structure of pyruvic acid. Some yeast and bacteria evolved procedures for converting pyruvic acid into other organic compounds. A large proportion, however, became acetyl coenzyme A, which was used to synthesize lipids and other components. As long as the environment remained free of oxygen, there was no means of oxidizing pyruvic acid farther to extract the great store of energy still bound in its chemical bonds.

Then between 1.8 and 0.6 billion years ago, the oxygen liberated from water by photosynthetic organisms continued to oxidize the earth's hydrosphere and atmosphere until the environment began to accumulate free oxygen. During this interim of 1.2 billion years some microorganisms evolved enzymes and components that became a series of reactions for the complete oxidation of pyruvic acid. It must have taken an extremely long time over countless generations for the genetic apparatus to evolve the developed pathway—but the selective advantage of the gain in energy was overwhelming.

We see today the aerobic breakdown of pyruvic acid in its most evolved state and know little of the various stages it must have gone through in its development. At least some of the reactions existed already as side reactions of fermentation to provide amino acids. These reactions were apparently modified and extended to create the aerobic pathway. Although it is a biochemical procedure that could not have been possible until atmospheric oxygen reached the Pasteur point of 1 percent of today's level, it probably was preceded by a long period of evolution. The pathway has come down in the plants and animals through the mitochondria, the sausage-shaped subcellular particles that are the suppliers of ATP in all eucaryotic cells.

The detailed chemistry of respiration was not discovered until the late 1930s, when Hans Krebs[3] pieced together a series of enzymic reactions involving tricarboxylic acids used to oxidize pyruvic acid to carbon dioxide and water. In the sequence, acetyl coenzyme A from the oxidative decarboxylation of pyruvic acid enters a mitochondrion where the acetate group couples with oxaloacetic acid to form the six-carbon compound citric acid. In the succeeding seven reactions, citric acid is degraded back to oxaloacetic acid through a procedure whereby the two-carbon acetyl unit is oxidized to two molecules of carbon dioxide by molecular oxygen. In the series the di- and tricarboxylic acids containing four, five, and six carbons are continually consumed and regenerated in the cyclic process. As a result, a single molecule of oxaloacetic acid can bring about the oxidation of an indefinite number of acetyl groups by being regenerated at the end of each cycle.

The membrane complex of the mitochondria, where the aerobic degradation of pyruvic acid takes place, is one of the most sophisticated structures ever evolved by biological systems. The molecular mechanisms of respiratory energy

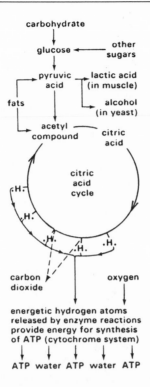

Figure 14.6. The metabolism of
carbohydrates through respiration
to produce ATP.

transformations are still not fully clarified, but mitochondria appear to be able to draw on it for ATP synthesis, transhydrogenation, ion transport, and other fundamental processes through a highly coordinated procedure. What is known is that structure and function are inextricably entwined where the enzyme complexes carry out the oxidation of pyruvic acid and fatty acids within the mitochondrial membrane.

For something as complex as the mitochondrion to have evolved, the selective advantage had to have been large. In examining the return of energy through respiration, it is easy to see what the mitochondria gained. All organisms contain the anaerobic pathway for glycolysis, inherited from our common procaryotic ancestors, but the eucaryotes with their mitochondria extended the degradation of glucose by oxidizing the pyruvic acid completely to carbon dioxide and water.

As long as biological systems were restricted to the anaerobic breakdown of glucose, they derived just two molecules of ATP for each molecule of glucose. This represented a yield of only 61 kilojoules of energy from a total potential of 2,870 kilojoules sealed in the chemical structure. The change of the earth's

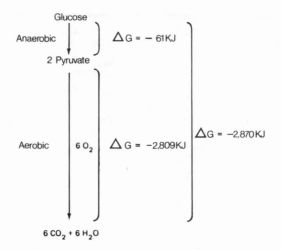

Figure 14.7. Comparison of energy yields of anaerobic degradation of glucose to pyruvic acid versus the oxidation of pyruvate to CO_2 and water.

atmosphere to an oxidized state with free oxygen permitted the degradation of glucose through respiration to draw from the untapped reserve. When the overall equation for the complete oxidation of glucose to carbon dioxide and water by aerobic organisms is written, it becomes

Glucose + 36 P_i + 36 ADP + 6 O_2 $\longrightarrow$ 6 CO_2 + 42 H_2O + 36 ATP

For nearly 3 billion years life remained on the microscopic level, but the breakout was slowly being evolved by the mitochondria, or rather their direct predecessors that developed the enzymic pathway for the complete oxidation of glucose. Evolution of the mechanism may have taken hundreds of millions of years, but the reward was immense. Instead of obtaining two molecules of ATP for each glucose as in fermentation, the organisms capable of oxidative phosphorylation obtained 36. It was this 18-fold bonus of energy that catapulted life out of the cellular confinement of the microbe into multicellularity and a whole new dimension.

15
The Question of Genesis

All organisms on earth, from the simplest cell to man, are machines of extraordinary powers, effortlessly transforming complex organic molecules, displaying elaborate behavioral patterns, and indefinitely producing from raw materials in the environment more or less identical copies of themselves. The life processes consist of exquisitely interlocked and controlled steps carried out in a marvellously architectural interior of the cell, with specialized regions where particular chemical reactions are performed. It has been estimated that a single human cell contains 100,000 enzyme molecules to mediate 1,000 to 2,000 chemical reactions, an average of 50 to 100 molecules for each process; and the human body contains 10 trillion cells, each serving a particular role in the whole organism.

The ability of enzymes to promote biochemical reactions is equally impressive. The rate of catalysis by most enzymes ranges from 1,000 to more than 500,000 molecules per minute. But the fastest rate observed seems to be by the enzyme catalase. One molecule of catalase can transform more than 5 million peroxide molecules in a minute.

Small wonder then that people have looked upon life as a phenomenon that could be explained only in terms of nonphysical laws. But as with all phenomena, a living cell has appeared inexplicable only because there had not yet existed sufficient information about the biochemical processes to piece together a logical sequence of events in its function.

At one time it appeared that, because life created order from disorder by its syntheses and growth, it defied the Second Law of Thermodynamics, which specifies that no processes can occur that increase the net order. This view, however, failed to take into account that the Second Law is valid only for closed systems. The universe as a whole, with no exchange of matter and energy from the outside, is steadily moving toward a state of complete randomness. Living systems, on the other hand, are open—not closed. It is the flow of sunlight from outside our environment that drives the life machine.

The impressive powers of synthesis by enzymes have been confounding simply because our technology generally has not reached the level of efficiency of life processes. The laws of physics and chemistry are derived by the statistical probability of molecules behaving in a defined manner. The kinetics of a chemical reaction are computed on the percentage of molecules of sufficient energy moving freely in a reaction mixture that collides with another appropriate molecule. The energy content of molecules in a population follows a distribution curve, and only those molecules with an energy above a particular level are likely to form a product. But biological systems have evolved a mechanism that is entirely different from the probability mechanism of chemistry and physics. Enzymes possess active sites where the donor and acceptor for chemical exchange are on the same molecule and positioned spatially so that the reaction occurs with near certainty when the substrate joins the enzyme. In this way the efficiency of the reaction is made maximal by dispensing with the inefficient procedure where the donor and acceptor molecules move independently in solution.

The living cell is thus capable of carrying out series of reactions with amazing efficiency. A single or small but highly organized group of molecules is held and oriented in regions of the cell so that a chemical conversion can be achieved through a chain of reactions with the probability of each reaction being essentially 100 percent. The collective behavior of molecules organized in this manner, except in living matter, was unprecedented in science until the recent development of solid-state physics.

The elucidation of the action of photosynthesis, metabolism, enzyme mechanics, the synthesis of proteins, and the replication and translation of nucleic acids has shown that they all follow established chemical and physical laws. There is no evidence of a vital force—nor is there any need to call upon the idea to understand the mechanics of life. Vitalism was never a principle that was thought out anyhow but was a sort of catchall concept to cover anything otherwise inexplicable.

Equally impressive has been life's complexity. There are approximately one million species of animals and a half million species of plants classified by biologists, with the number of undescribed living species estimated to range up to 10 million. The number of extinct species represented by fossils may be even larger. Life has become enormously diverse, expanding into virtually every accessible ecological niche from oxygen-depleted oceanic oozes and ammonia-rich soils to mineral deposits with high radioactivity content. There are microorganisms that live in pools at Yellowstone National Park at temperatures of 80°C (176°F), and there are microflora in the Don Juan Pond in Antarctica that metabolize in calcium chloride water at temperatures as low as -23°C (-9°F). Other bacteria, algae, and fungi can live in extremely acidic or alkaline environments, and much to the annoyance of nuclear physicists, the bacterium *Pseudomonas radiodurans* continues to thrive in the large neutron flux at the cores of swimming pool atomic reactors.

We tend to be amazed at this immense variety of life on earth, but beneath the diverse exterior there lies a great commonality. The advances of biochemistry have revealed that all biological systems use basically the same processes to function as self-sustaining organisms. The nucleic acids and proteins are absolutely central to living processes, and the subunits for these biopolymers are the same for all organisms. Even the stereochemistry, the left- or right-handedness of biological molecules, is the same. There is also one principal class of compounds used by all forms of life to transfer chemical energy, and these agents are the nucleotide phosphates. The porphyrins, the cyclic compounds that form the active nucleus of hemoglobin, chlorophyll, and the cytochromes, are ubiquitous constituents. But the most impressive of all is the similarity in translating the nucleic acid information to protein structure. Since the genetic code was broken in the 1960s, it appears that the language of heredity is the same for all living things. These commonalities and many others, particularly where no obvious selective advantage exists, strongly imply that all forms of life on earth are descendants of a single common ancestor.

Since life evolved from a primitive microscopic predecessor, it has become immensely complex. The weight of the cellular DNA in mammals has been found to be 6.5×10^{-12} gram. If one nucleotide pair weighs 1.03×10^{-21} gram, then the amount of DNA in each mammalian cell is equivalent to 3.2 billion nucleotide pairs. Considering an average protein to be 500 amino acids long, which corresponds to a gene of 1,500 nucleotide pairs, then each mammal has 2.1 million genes.[1] R. J. Britten and D. E. Kohne[2] have suggested that 40 percent of the DNA consists of sequences repeated between 10,000 and 1,000,000 times. These are long stretches of DNA that do not code for a sequence of amino acids or RNA molecules but serve to separate the information segments of the molecule. After subtracting these repetitions, there still remains space for as many as 1,250,000 different genes in the chromosomes of each species of mammal. In what manner could such complexity have evolved in a controlled procedure in 3.5 billion years?

A parallel complexity followed the development of multicellular systems. When unicellular organisms evolved to multicellularity, a single cell was able to give rise to an astronomical number of duplicates. The cells that form the human body are derived from a single fertilized egg that divided into 2, then 4, 8, 16 . . . and on by geometric progression to form the whole person. As incredible as it may seem, it takes only 41 divisions for a single cell to become 10 trillion.

Geometric progression is an extremely effective means of attaining large numbers, and molecular evolutionists have discovered that the principal mechanism in evolution is the doubling of the gene through mutation. With each doubling, a single amino acid substitution in one copy can add a new protein. In this way proteins as variant in function as lactalbumin from milk and the enzyme lysozyme

Table 15.1. Hypothetical doubling of a gene at
188-million-year intervals.

N^2	Time (millions of years)	Number of genes
21	0	2,097,152
20	188	1,048,576
19	376	524,288
18	564	262,144
17	752	131,072
16	940	65,536
15	1,128	37,768
14	1,316	16,348
13	1,504	8,192
12	1,692	4,096
11	1,880	2,048
10	2,068	1,024
9	2,256	512
8	2,444	256
7	2,632	128
6	2,820	64
5	3,008	32
4	3,196	16
3	3,384	8
2	3,572	4
1	3,760	2
		1

have evolved from the same gene. The doubling of the genetic material would
have been a rare event. Nevertheless, for a single gene to become 1.6 million
through geometric progression would have required doubling only 21 times since
the beginning of life on earth, or at an average interval of 188 million years.

The path to complexity was not as straightforward as the geometric progression
of a single gene. There have been other features involved in mutation and
evolution, and presumably the primal cell consisted of a multitude of small
polymeric substances, rather than a single gene. Moreover, the time scale is
uncertain because genes evolved not to maximum size but to optimal size for
the enzymes. The duplication of a large genome (all the combined genes) would
signify extensive duplication of cellular material by multiple copies of similar
proteins, introducing conditions that may have been more advantageous to early
life than to later forms. Nonetheless, the doubling of genes with subsequent
accumulation of point mutations apparently was the most influential and effective
means of attaining size and diversity, and certainly some ancient duplications
have involved the whole genome.

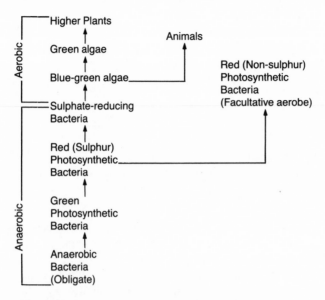

Figure 15.1. Evolutionary development of ferredoxins.

The amount of DNA measured from each of a large variety of mammals has been found to be nearly the same. Within mammalian lines there has been duplication of some small segments of chromosomes, as indicated by the beta and delta chains of hemoglobin, but no duplication of the entire DNA. Although mammals may have nearly the same amount of DNA, as a class, however, they have about four times the amount found in protochordates and some fishes. It appears that there was duplication of the whole genome twice in the course of evolution from protochordates to mammals. The first occurred sometime around 465 million years ago, when early vertebrates evolved from the primitive chordates. The second genome duplication occurred during early reptilian or pre-reptilian evolution about 320 million years ago.

The case of ferredoxin clearly illustrates the manner in which an enzyme evolved from a simpler structure. Ferredoxins are nonheme, iron-containing proteins found in anaerobic and photosynthetic bacteria, blue-green and green algae, as well as the higher plants. They participate as electron carriers in a wide variety of biochemical processes, including photosynthesis, nitrogen fixation, sulfate reduction, hydrogenase reactions, and other oxidation–reduction reactions.

The polypeptide chains of plant ferredoxins at 55 amino acid residues are nearly twice as long as those from anaerobic bacteria, whereas the ferredoxin of *Chromatium,* a photosynthetic bacterium, is intermediate. Furthermore, the two halves of bacterial ferredoxins are clearly related. It appears that all ferredoxins are derived from the repeated duplication of a primordial sequence of four amino acids, combined with deletions and amino acid substitutions. There

Table 15.2. Composition of *Dialister pneumosintes*.

	(Grams × 10^{-14})	(Daltons × 10^8)	Percent
Dry Weight	2.80	160	
DNA	0.13	7.8	4.66
RNA	0.30	18	10.33
Protein	1.20	72	43.00
Carbohydrate	0.47	28	16.45
Lipid	0.16	37	21.70

Note: A weight of 0.13 × 10^4 grams of DNA is equivalent to 6.5 × 10^8 daltons (one dalton is the molecular weight of one, or 1 gram is approximately 6 × 10^{23} daltons).

was a doubling of the gene length in the bacterial line which occurred just before the divergence of *Chromatium,* then an additional doubling independent of the bacterial doubling that led to the plant ferredoxins.[3]

Life has evolved from the simple to the complex, pulled by the competitive advantage of the most efficient organism or biochemical procedure to appear through mutation. For this reason, we are tempted to look to the simplest organisms as models of early life. There is value in this approach because there exist today living representatives of many of the stages of evolution throughout life's long advance. This approach must be taken with caution, however, for all contemporary life has an equally long ancestry. Although microorganisms appear to be simpler forms of life, they are not a primitive form of life. The enzyme structures and metabolic processes that bacteria share with us are no less sophisticated and efficient than our own. We differ because we have each evolved to exploit different ecological niches.

Though their polymeric constituents may be of equal refinement, the cellular architecture of procaryotes is considerably simpler than that of eucaryotic cells, and there are other features that give sufficient grounds to believe that procaryotes preceded the latter. To put things in perspective, we should examine the composition of a procaryote. The bacterium *Dialister pneumosintes,* for example, has the dimensions of 0.4–0.6 micrometer wide and 0.5–1.0 micrometer long. Since a micrometer equals 1/1,000 millimeter, or 1/40,000 inch, the *D. pneumosintes* is so small that 250,000 of them could be placed on the period at the end of this sentence. Nevertheless, it is a simple matter to measure its chemical composition and calculate its constitution.

Much can be deduced from the organism's composition. Since a double-stranded DNA must have approximately 20 times the weight of protein for which it is supplying the information, 6.5 × 10^8 atomic mass units of DNA will carry the information for 3.3 × 10^7 atomic mass units of protein. Assuming an average molecular weight for protein to be 40,000, there then must be 800 molecules of different proteins in a cell of *D. pneumosintes.* If all the proteins are enzymes, for a single molecule per reaction, the cell uses 800 reactions to be an autonomous functional system. The bacterium, therefore, is probably using several hundred reactions, which seems like a reasonable figure since in the larger *E. coli,* the microorganism most extensively studied, there are already 500 known biochemical reactions.[4]

Small as they may seem, these are not the smallest autonomous cells. The smallest free-living cells known are the pleuropneumonia-like organisms (PPLO) called mycoplasmas. These organisms cause serious disease in sheep, goats, chickens, and turkeys and are found as harmless saprophytes on the mucous membranes of animals and man, as well as in sewage and the soil. While an amoeba has a mass of 5×10^{-7} gram, a PPLO weighs 5×10^{-16} gram and is only about 0.1 millimicron across. It can be seen only with an electron microscope.

Because of its size, a PPLO has room for only about 100 enzymes. These organisms grow very slowly. Apparently they live in an environment where they receive free many necessary constituents that larger organisms have to synthesize. Nevertheless, the PPLO are fully functional reproducing organisms that survive on no more, and probably considerably fewer, than 100 biochemical reactions. Were there an environment in which all the necessary building blocks and such energy sources as ATP were provided, a functional organism could be substantially smaller and simpler than even the PPLO. In fact, the inside of a cell is such an environment and that is why viruses are smaller.

Viruses, however, are not self-sustaining organisms and cannot be regarded as a form of life. They should not be considered primitive but rather a sophisticated form of parasitism, for they too are products of 3.5 billion years of evolution. Viruses do not appear to be the end product of evolution from a simpler structure but are either an aberrant derivative of a cell or a degenerate product from a higher form of life.

Viruses are the simplest organism,* but there is another type that exists in the realm between bacteria and viruses. These are the rickettsiae, which, like the viruses, are parasites that multiply only in a host cell but are nearer to bacteria in size, complexity, and metabolic capacity. Rickettsiae are responsible for several diseases, including epidemic typhus and Rocky Mountain spotted fever. An interesting feature of rickettsiae is their similarity to mitochondria. Both have the components for the Krebs cycle of reactions, both use NADP, and neither can carry on its usual oxidations after having been frozen in a salt medium, but can be protected against the loss of activity by the addition of glucose.

One theory is that viruses originated as degenerate genes that broke from the cell, a concept prompted by the extreme specificity of the viruses. A more attractive hypothesis is that first the rickettsiae and then the viruses evolved from bacteria, and as they developed a parasitic mode of existence they lost as su-

---

* At around 10^{-18} atomic mass units, viruses are about 100 billionth the size of an amoeba. The amoeba, therefore, is as far removed in size from the viruses as we are from the amoeba.

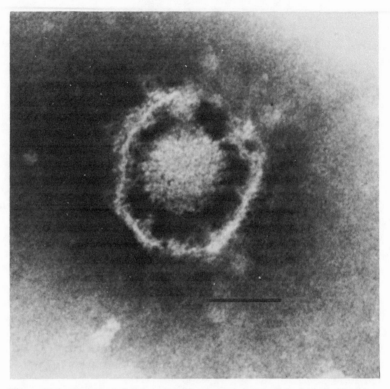

Figure 15.2. Electron photomicrograph of a herpes virus particle (virion). The virus envelope surrounds the nucleocapsid that has the shape of an icosahedron. The capsomeres (morphological) subunits of the capsid are apparent. The bar represents 100 nm; magnification is 200,000 X.

perfluous the capability of synthesizing some enzymes necessary for a free-living organism.

Men have marveled at the beauty and efficiency of plants and animals and have expressed the idea that such developments must be singular events that could never have arisen through natural processes alone. But nature is not based on the improbable. Various life forms, their structures and functional parts, evolved as the most advantageous of a large number of possibilities, which means that the selective process made their development probable under the circumstances. The streamlined shape for high-speed marine motion appeared with the *Stenopterygius* and other Mesozoic reptiles, with the tuna, which is a fish, and with the dolphin, a mammal. That this shape evolved independently three times in unrelated life forms does not indicate a rare coincidence, but rather than hydrodynamics has a narrow range of solutions to the problem. The eye, which has been regarded as a wondrous organ, developed independently several times, indicating that such a structure is the best solution to visual recording.

Calculations that show that a chance assembly of the base triplets of a nucleic acid would be of 10^{87} different ways,[5] or that the probability of a protein of a molecular weight of 60,000 to form by mass action is 1 in 10^{605} are meaningless. The formation of life was neither by chance nor improbable. Such calculations are designed to prove that life could not have happened through natural circumstances. But they are based on the incongruous premise that life sprouted in full bloom with all the complexity and refinement of contemporary organisms. Life began simply and slowly and evolved as the result of a systematic survival of combinations of substances that worked and the rejection of all that did not.

Complex systems, from empires to biological structures, do not appear suddenly in their full stages of development. To the contrary, they have to grow by stages, with each level established as a stable platform before the next higher tier is undertaken. Life developed by tiers of chemical structures, each tier progressively larger and more complex than the previous one, the units of each tier having within their chemical nature the means of creating the succeeding stage of development.

Nucleotides condensed to polynucleotides, amino acids condensed to polypeptides, macromolecules assembled into a functional cell, and the tier of cells in turn became associated into multicellular plants and animals. At each level of this pyramid the units had to undergo self-assembly to form the next higher stage. But here lies a crucial feature of genesis. If one dissociates a contemporary living cell and then recombines all the biological constituents, they do not reassemble into a functional cell. Nor can one dissociate an advanced multicellular plant or animal into its constituent cells and reconstruct the original organism. At this stage of development the feature responsible for self-assembly appears to have been lost.

At first glance this seems to refute the basic principle just when it applies to the level at which life formed. But we saw that multicellular organisms at the early stages of evolution—sponges, coelenterates, and embryos in their initial development—do possess powers of self-assembly. It was only after multicellular organisms had evolved further that the organization and control of these various cells were built into the overall genetic blueprint and the cells became too specific for self-assembly.

What is happening is that the more complex a unit becomes, the more specific are its intra- and intermolecular associations, and hence the less probable for the unit to reassemble correctly, once dissociated. The denaturation of enzymes is an excellent example. Small polypeptides, like nonpolymeric chemicals, have a relatively stable and uniform spatial arrangement. Proteins, on the other hand, occupy space as complex and often fragile structural configurations dictated by the amino acid sequences and a multiplicity of internal interactions.

In the biosynthesis of a protein, as the polypeptide chain grows from the ribosome, it coils, twists, and bends into its shape and is held by the internal interactions as they form with the emerging protein. Under the chemical environment of the cell, the three-dimensional configuration remains. If, however, the protein is isolated and subjected to even modest changes in the pH or salt concentration, as well as elevated temperatures, the weak forces holding the protein in its configuration rupture. The polypeptide chains reform intra- and intermolecular bonding, but they have lost their uniformity because now a completed chain has many possible ways of interacting. The probability of the protein assuming its original configuration is generally quite small and the protein is said to be denatured. When the protein is an enzyme whose activity is dependent upon its specific configuration, denaturation results in the loss of enzymic activity.[6]

When the components and the architecture of a biological system are small and simple, the probability of self-assembly is the greatest. For instance, ribosomes have the capability of self-assembly, and some viruses have been dissociated and recombined to yield several percent of reconstituted organisms.[7] The chemical nature of many biological compounds is to aggregate in fairly specific supramolecular combinations—the base pairing of complementary nucleotides is an example of such aggregating. There are, nevertheless, additional interactions, including other base pairings, that are possible but not used by biological systems. Some of these may have been a part of primal cells but were dropped as being less useful than those adopted by organisms. Without a doubt, though, the most dramatic form of self-assembly is the spontaneous impulse of lipids toward structure formation.

The minimum requirement for the formation of the primal cell was that the associating components had to be chemically adequate and appropriately organized to carry on regeneration of the cell. Probability would have favored those cells that had the smallest and the fewest number of components while still being functional. The biochemical transformations would have been slow and inefficient, but there was no competition and the only enemy was dissolution. The threshold that the initial living systems had to cross was to be able to grow and reproduce at a rate faster than hydrolysis. Once across the hump, the cell accelerated by mutating to ever greater efficiency—and the descendants of this first successful form of life inherited the earth.

As testimony to life's simple beginning, the complexity of advanced organisms is based on a surprisingly few number of substances. Of the billions of possible organic compounds, fewer than 1,500 are employed by contemporary life on earth, and these 1,500 are constructed from fewer than 50 simple molecular building blocks.

16
The Essentials of Life

Does life have a unique chemistry that was met by singular circumstances over 3.5 billion years ago on primordial earth or was life the inevitable consequence of ordinary geological conditions? In order to answer this question, we need to delineate the components that would have been necessary for a functional living cell to have formed. By establishing the absolute requirements for a biological system, we can then understand what specific conditions were necessary for the origin of life on earth.

An examination of the chemical composition of living things shows only common elements. Of the 92 naturally occurring elements, 40 are found in plants and animals, but only 18 are commonly required. Of 36 elements found in protoplasm, 4 actually make up 98 percent of the total composition. These four are carbon, hydrogen, nitrogen, and oxygen. There are small quantities of phosphorus, sulfur, sodium, chlorine, calcium, magnesium, and potassium; and the trace metals, iron, boron, molybdenum, copper, cobalt, zinc, and manganese represent less than 1 percent. There are no rare or unusual elements among them.

Carbon, hydrogen, nitrogen, and oxygen, which predominate in biological systems, are, except for helium, the four most abundant elements in the universe. Not only are carbon, hydrogen, nitrogen, and oxygen among those elements with the highest cosmic appearance, they also have chemical properties so distinct from the others that they are unique in their roles as essential elements. They are of low atomic weights and form a number of common stable and volatile compounds. For this reason, these four elements became concentrated in the gaseous and watery envelope that sheaths the earth's surface.

Table 16.1. Volcanic gases.

Steam	H_2O	
Carbon dioxide	CO_2	
Nitrogen	N_2	
Sulfur dioxide	SO_2	
Hydrogen	H_2	
Carbon monoxide	CO	Decreasing
Sulfur	S	Amounts
Chlorine	Cl_2	
Hydrogen sulfide	H_2S	
Hydrochloric acid and other acids	HCl	
Volatile chlorides of iron, potassium, and other metals		

Life arose from the mass of volatiles outgassed from the earth's interior that condensed and accumulated to form the atmosphere and the oceans. Without the geological evolution of the earth, life could never have formed on its surface. Not only does the earth revolve around the sun in a narrow zone with margins of only a few percent to permit the conditions that allowed life to develop, but if the earth had been appreciably smaller, like the moon, any atmosphere would have escaped from the smaller gravity. But most of all, with the greater surface to volume ratio, the heat from the trapped nuclides would have been dissipated to outer space without the buildup to volcanism. The zone of magma would not have formed, the earth's interior would not have differentiated into concentric layers, and the excessive volatiles would not have been outgassed to the surface to create the atmosphere and oceans. Without these, the earth would never have spawned life.

Carbon has the indispensable property of being able to form chemical bonds, not only with a large number of other elements, but also with other carbon atoms. Some other elements share this feature of concatenation, though only to a limited extent. Characteristic of carbon is its ability to link into long sequences with single, double, and triple bonds and form ring compounds or enter into a vast number of combinations with other elements. The number of known carbon compounds is now estimated to be over one and a half million. It is this immense variety of carbon architecture that has furnished the necessary complexity for the interplay of chemical and physical properties that makes a self-sustaining biochemical entity.

One of the features of living systems that distinguishes them from the inanimate is their ability to reproduce. With the fundamental nature of a cell being chemical and physical structuring of the constituents, we are dealing primarily with molecules that replicate spontaneously. The manner in which nucleic acids are duplicated and how the sequence of the four nucleotides represents a coded message that can be translated to a chain of amino acids in the synthesis of proteins have been shown. The basis of life is that uniqueness of nucleic acids to be copied.

The primary role of nucleic acids is to serve as a relatively stable blueprint to the organisms' construction. For this function it has to be large enough to carry the code for the structure of a peptide. Since 3 nucleotides are needed for each amino acid, the size of the postulated elementary proteins would dictate the nucleic acid requirement. If ferredoxin began with a peptide 4 amino acids long, as is now believed, the corresponding nucleic acid would therefore have been at least 12 subunits in length.

But a second role of nucleic acids is as transfer RNAs. In order to copy the coded nucleic acid and translate its composition to a sequence of amino acids, a series of tRNA molecules would be required. In contemporary organisms, there is a tRNA for each of the 20 amino acids.

In 1965, Robert Holley and his coworkers[1] determined that the nucleotide sequence for alanine tRNA is a polynucleotide of 75 subunits that could fold back upon itself and be held in a cloverleaf arrangement by hydrogen bonding between opposing complementary bases. Since then, the sequence of some 75 different tRNA molecules has been determined. The number of subunits ranges from 74 to 91, and they all form the same general cloverleaf configuration.[2]

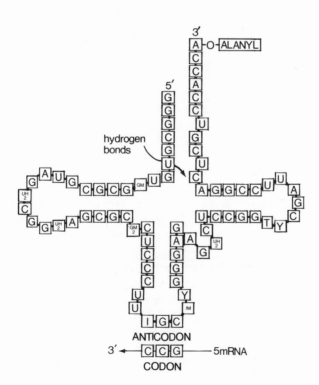

Figure 16.1. The cloverleaf configuration of transfer RNAs (alanine tRNA).

The maximum size required for polymerized nucleotides to function as tRNAs would appear to be not the full cloverleaf arrangement but at least the chain folded back on itself to form the hairpin loop to allow for the anticodon triplet at the bend. Since the amino acid specificity of the tRNAs would have evolved early, the stems of the initial tRNAs requiring the side-loop of the cloverleaf were probably nearly as long as they are now, which would place a minimum size for the nucleotide chain at about 40 units.

Replication of nucleic acids would have preceded protein biosynthesis in the development of life, and because of the high specificity and striking closeness in the structure of tRNAs, they are probably all derived from a single ancestral molecule.

That macromolecules had to precede the first living cell is easy to see. But if one looks at the mechanism by which nucleic acids and enzymes are formed in contemporary organisms, he faces a dilemma: nucleic acids carry the information and mechanism for synthesizing both themselves and the enzymes, but enzymes are the catalysts for the reactions to prepare both. How could either have been formed without the presence of both?

This question has puzzled and divided scientists into those who believe nucleic acids came first[3] and those who contend that enzymes or enzymelike macromolecules would have been the more likely original components.[4] In order to reconstruct a rational origin of a biological system, one needs either to show that enzymes (or suitable substitutes) could have been produced abiotically or be able to explain how a biological system could begin without them.

The fundamental basis of living things is to grow and maintain themselves by building molecules. They synthesize and compact macromolecules from smaller units—a process that requires an input of energy. The free energy liberated in the hydrolysis of a peptide link of a protein molecule is between 2 and 4 kilocalories per mole; which is to say, to join each amino acid to the peptide chain in synthesizing the protein requires 2 to 4 kilocalories. A similar requirement exists for polymerizing nucleotides.

One way of meeting this energy demand to form chemical bonds would be to raise the temperature above the boiling point of water and high enough that the thermal energy of the reactants exceeds the energy needed to form the peptide linkage. But as molecules become more complex for a biological system, weaker interactions play a major role in maintaining the ascending architecture, and the structure becomes more susceptible to the disruptive action of thermal energy as it exceeds the chemical energy of these bonds.

For these reasons, organisms have developed a highly selective and controlled input of energy to fuel their processes. In order to convert the uphill reactions of synthesis into spontaneous downhill reactions, biological systems activate one of the reactants to a high-energy derivative. To do this requires a third reactant of an even higher energy potential to pass some of its energy down to the reactant for activation.

The bioenergetics of a living cell is essentially like the use of an electric power plant. Heat energy from burning coal is converted to electricity by a generator and carried to homes, where it is converted to heat and motion. In the process, more energy is required to generate the electricity than is ever recovered. In the same manner, a plant cell traps energy of sunlight, transfers it in special compounds to other parts of the cell, and recovers the energy to build chemical bonds.

Phosphoric acid derivatives are particularly feasible for this role of holding in their structure the energy collected from sunlight. It is the stored chemical energy of the pyrophosphate bond that fuels biological syntheses. Consequently, these high-energy compounds are essential components of all biological systems. Although other pyrophosphate derivates could have played an initial role, the stablility advantage of ATP suggests that it may have been the original constituent of life.

The requirement of pyrophosphate derivatives makes phosphorus an essential element. But of the six most common elements required for a biological system—carbon, hydrogen, nitrogen, oxygen, sulfur, and phosphorus—only phosphorus does not have a stable volatile derivative—and there is no substitute. Arsenic is more closely related to phosphorus than any other element, but the acid anhydrides of arsenic are not stable in water, and it is the aqueous stability of high-energy ATP that makes it possible for it to be used in the synthesis of nucleic acids, proteins, and other cellular components.

This introduces an interesting inconsistency. The presence of phosphates in seawater is in only micromolecular quantities, making it one of the scarcest elements in the ocean. Asimov[5] compared the composition of seawater to the required elements to determine what factor restricts the growth of life. Taking the tiny crustacean copepod as representative of ocean animal life, he compared the percent of essential elements in the copepod's composition to that of seawater. Any higher occurrence in the copepod indicated concentrating of the element by the animal. Four elements—carbon, nitrogen, phosphorus, and iron—had large concentration factors. Carbon and nitrogen are available from the atmosphere through the food chain; but for the other two, the concentration factor was four times greater for phosphorus than for iron. Asimov concluded that the limiting factor to the growth of life in the ocean is the supply of phosphorus. This is evident in inland waters where phosphorus pollution from detergents and fertilizers sets off a rapid growth of algae.

But as critical as phosphorus may be for life, it is not a rare element. Certain rocks, particularly granite permagtites, frequently contain large deposits of the tricalcium phosphate minerals apatite, $Ca_3(PO_4)_2 \cdot CaCl_2$, and fluorapatite, $3Ca_3(PO_4)_2 \cdot CaF_2$. The difficulty is that the calcium salts of phosporic acid are extremely insoluble, effectively removing phosphorus from any waters in which the calcium concentration exceeds that of phosphorus. It is quite possible that

Table 16.2. Concentrations of frequently required heavy metals.

	Plant requirement	(mM) Crust	Sea and air
Iron	0.400	900	0.0001
Manganese	0.200	17	0.000036
Zinc	0.060	1.1	0.000075
Copper	0.020	0.87	0.000050
Cobalt	0.0003	0.43	0.000002
Molybdenum	0.0002	0.015	0.0001

Source: Adapted from J. H. McClendon, J. Mol. Evol. *8*, 175–195 (1976).

life may have originated in close association with phosphorus-bearing rocks or in sedimentary deposits of secondary phosphates washed from such rocks.

John McClendon[6] of the University of Nebraska made a comparison of the abundances and the requirements of the elements found in biological systems. He found that nine elements, because of their unique requirements, would have been essential for the origin of life. These are carbon, hydrogen, nitrogen, oxygen, phosphorus, sulfur, magnesium, potassium, and iron. Carbon, hydrogen, nitrogen, and oxygen would have been abundant in the gaseous components of the atmosphere, phosophorus was concentrated in rock deposits, and the remaining required elements would have been in seawater under the reducing conditions of primordial earth.

The heavy metals, iron, manganese, zinc, copper, cobalt, and molybdenum, are predominantly in the earth's crust—the oceans are nearly depleted of them. Nevertheless, they are essential elements for most plants and animals. McClendon found that the concentration of essential heavy metals in plants did not correlate with oceanic concentrations but did correlate with their occurrences in the crust. This would be consistent with plants, which colonized the earth's land areas before animals, adjusting their biological composition to the available crustal constituents, the incorporation of metals suitable for particular enzymes being affected by the abundance of the metal.

But did the heavy metals have any role in the biochemistry of the early forms of life on earth? The answer depends upon the order of events in the development of primal life. Iron, copper, and manganese function primarily as components of proteins used in photosynthesis. It has been suggested by D. O. Hall[7] that ferredoxin, the iron-containing protein involved in the transfer of electrons in the photosynthesis series, may have been the earliest protein formed, so it is quite conceivable that a means of harnessing the energy of sunlight arose early in the development of biological systems. If this is so, in order to explain the evolvement of photosynthesis by organisms in the ocean we must account for a great availability of these elements than there is in seawater today. We now can see that when photosynthetic organisms arose over 3 billion years ago, the chemistry of these elements under the reducing conditions of Archean earth would have been different from what it is at present.

Many of the heavy metals, being more abundant in the crust than in seawater, even under reducing conditions, probably became incorporated into enzymes of advanced biological systems after evolutionary development created a need. Cobalt and molybdenum appear to be of this category. Cobalt as a part of vitamin B_{12} and molybdenum in xanthine oxidase for vertebrates are clearly of late development. But cobalt is also associated with nitrogen fixation, as is molybdenum with nitrogenase. Nevertheless, legumes, with their symbiotic bacteria for fixing nitrogen, no longer need molybdenum if ammonia is available.[8] Margulis[9] holds the contrary view that nitrogenase in nitrogen fixation began in early organisms; and because the essential element molybdenum is in low abundance, some investigators have suggested an extraterrestrial origin of life.[10] It seems more probable that neither cobalt nor molybdenum was essential when the first biological systems appeared on earth but became incorporated after considerable evolution had taken place.

The role of magnesium as a part of chlorophyll would have developed with the advent of the blue-green algae at a later time. The fact that vanadium and nickel porphyrins are found in petroleum is not an indication that these metals were used by early plants in photosynthesis. A more plausible explanation is that they resulted from the magnesium being displaced to form the more stable vanadium and nickel complexes during the formation of petroleum.

Calcium, on the other hand, which is so essential in contemporary animals for many physiological reactions, as well as for skeletal structure, may not have been a primitive essential element. It seems that it came to be a derived requirement when plants and animals evolved extracellular structures.

In his examination, McClendon found that there is an absolute limit below which an element may not fall and still be available in adequate quantities for biological systems. The minimum concentration is 1 to 2 nanomolar in the ocean and 10 to 21 micromoles per kilogram in the earth's crust. By this criterion the elements that are indispensable for life are in abundance. Only cobalt and molybdenum are near the lower level of occurrence, but dependence on them was apparently acquired late and isn't absolute. Some elements, like rubidium and strontium, are sufficiently abundant and chemically suitable for biotic processes but are not used because of a greater supply of a similar element.

There are two additional substances that are indispensable for the formation of biological systems: water and lipids. Each has unique properties different from the other, and it is this opposing behavior of their respective properties that makes them absolute requirements for life.

Water is more than a solvent for biochemical reactions; in most cases it is a reactant. Cellular chemistry involves principally the formation and hydrolysis of acid derivatives: the phosphorylation and the coupling of nucleotides are esterification reactions; the synthesis of proteins is amidation. Since these condensations involve the elimination of water between the two reactants, and hydrolysis is the addition of a water molecule, water is directly a part of the reactions. When these reactions are carried out in an aqueous medium, the concentration of water is 55 molar. At such a high concentration, it plays an important role in establishing the position of equilibrium of the reaction.

Because water is amphiprotic, it can act as either an acid or a base, and it serves in both roles in biological processes. It acts as a weak base and accepts a proton from the dissociation of acid carboxylic groups.

$$\underset{\text{acid}}{\text{R-COOH}} + \underset{\text{base}}{\text{H}_2\text{O}} \rightleftharpoons \underset{\text{base}}{\text{R-COO}^-} + \underset{\text{acid}}{\text{H}_3\text{O}^+}$$

And in the protonation of ammonia it is an acid; that is, a proton donor.

$$\underset{\text{base}}{\text{NH}_3} + \underset{\text{acid}}{\text{H}_2\text{O}} \rightleftharpoons \underset{\text{acid}}{\text{NH}_4^+} + \underset{\text{base}}{\text{OH}^-}$$

In the catalysis of mutarotation it acts as *both* an acid and a base.

But the fact that water is a liquid at all in an ordinary temperature range is due to one of its most outstanding features. Water does not exist as single molecules under normal circumstances, but rather it is an intermolecular array of molecules held together by hydrogen bonds. A molecule of water can participate in the formation of a maximum of four hydrogen bonds. It is this network of binding that gives water a high boiling point for so simple a molecule. If H_2O had properties more like those of the hydrides of the other elements of Group VIa of the periodic table, H_2S, H_2Se, and H_2Te, its boiling point would not be 100°C—it would be −81°C (−112°F).

Water is one of the few substances that expands when it freezes, making the solid state less dense than the liquid. If it behaved like most materials, then ice, instead of floating, would sink, lakes would freeze solid during winter, and as ice accumulated on the bottom of the oceans, the volume of water would become smaller and increasingly saltier. Another character of water is its unusually high heat capacity, which acts as a stabilizer for the internal biochemical environment. It is conceivable that life could have existed without these special features of water, although being without them would certainly have created a whole new set of problems.

But these features are related to the one property of water that makes it indispensable for life. The ability of water to form an array of strong hydrogen bonds with itself is a property that has great influence on biological systems. Much of the support that holds complex proteins in their tertiary arrangement is from the associated structure of water around them. It is a particular relationship between water and lipids, however, that has created the familiar features of cells. By being immiscible, water and lipids have an interface that permits a zone of contact for reaction between dissimilar substances which would not be able ordinarily to interact in a single phase system and makes catalysis possible.

Lipids form a class of organic compounds that are insoluble or sparingly soluble in water, but are soluble in lipophilic organic solvents. Fatty acids, triglycerides, phospholipids, and glycolipids belong to this class of compounds. Lipids that contain paraffinic chains such as fatty acids or their derivatives become insoluble when the aliphatic chain exceeds about four methylene units. At this point the water lattice extending from the polar end of the lipid molecule can no longer encapsulate the hydrocarbon tail and the lipid is insoluble. And since they are hydrophobic, the lipids are excluded from the water structure and they coalesce. It is this physical behavior that is vital for the formation of a cell.

A cell is a compact arrangement of biochemical molecules enveloped in a semipermeable membrane. The membrane is a bilayer of phospholipids with proteins closely associated with it in various arrangements, including attachment to the polar ends of the lipids at the water–lipid interface. But because cells exist in aqueous environments, they have several features that are satisfied only by lipids.

A cell is insoluble. Even substances that are extremely insoluble will dissolve until they saturate their surrounding medium. For a cell with a membrane short of being completely insoluble, the effect would be devastating. It would continually be facing termination by dissolution.

In addition to being insoluble, the membrane has to be hydrophobic, for it is the exclusion from water that causes lipids to come together to create a bilayer that can act as the plane of contact between separate aqueous compartments. Moreover, lipids are lighter than water, whereas other biological compounds are slightly denser. When lipids are combined with other compounds, the density of the aggregate can be near that of water.

Phospholipids are fundamental and universal with biological systems. Having a polar end which interacts with water molecules and hydrocarbon chains that are excluded from water, phospholipids in water automatically orient in a definite configuration. When a mixture of phospholipids and water is agitated, liposomes—small, spherical particles—are formed in which the phospholipid molecules are aligned in double layers with the hydrocarbon chains coalesced and the polar phosphate groups extended outward in association with the water. Liposomes may be formed in a series of concentric bilayers, each oriented in this fashion. It is a form of self-assembly that creates an encapsulation of substances inside the sphere and restricts diffusion of ions and molecules across a lipid barrier. It forms spontaneously with phospholipids in water—it is a feature of all biological systems.

The minimum requirements for a functional cell appear to be nine essential elements, a proper aqueous environment, and the formation of small nucleic acids, ATP, peptides, and phospholipids. Nucleic acids are composed of purines, pyrimidines, and ribose moieties, peptides of amino acids, and phospholipids of fatty, glycerol, and several other small subunits. These are the building blocks of the macromolecules and structures of a living organism.

But the problem is that all these building blocks for the development of a living organism come from one source—other living organisms. We find ourselves back at the beginning. If the building blocks for constructing a living system come only from living systems, how did the first cells come into being?

17
The Search for the Building Blocks

In 1807 when Berzelius defined organic compounds as the exclusive products of biological systems, he was more correct than not. The synthesis of urea from ammonium cyanate by his student Friedrich Wöhler in 1828 only demonstrated that organic compounds could be synthesized from inorganic chemicals. Without the directing hand of man, organic compounds remained within the domain of biology. Organic chemistry developed as a branch separate from the inorganic, in no way disputing the fact that biological compounds are produced in nature only by living organisms.

But then the principle of spontaneous generation, the concept that microbial life rose spontaneously, was finally put to rest by Pasteur in 1867 by a few simple experiments. Even the belief that life was due to a vital force that set it apart from the inanimate was losing adherents as the chemistry of biological systems became elucidated. This discrediting of the old principles before they were replaced by new, however, created a severe gap in science's understanding of the physical world. The origin of a primal cell remained an unsolved mystery. If all living things can be defined in terms of a biological cell, and the cell is a functional unit of biochemical compounds and processes, how then did the first cells come into being on primordial earth?

In particular, how could the first living cells have obtained amino acids to make their proteins? Once cells synthesized proteins they could produce enzymes which were the key to the entire biological machinery. But all amino acids found in nature come from a plant, animal, or microbe at some time or other. Even for plants to use carbon dioxide, minerals, and water to produce amino acids, sugars, and lipids they need a complex array of enzymes and cellular structures. With the barren geological conditions of prebiological earth, it was inconceivable how even the simplest type of primitive cell began. It was a problem that was to perplex scientists for over a century.

One of the earliest attempts to formulate an answer to the enigma was made by the eminent German physiologist Eduard Pflüger. In a remarkable paper published in 1875, Pflüger[1] postulated that simple organic compounds may have formed from inorganic minerals by natural conditions of early earth. Drawing from chemical knowledge of one hundred years ago, he pointed out that potassium and carbon heated to a high temperature together in air gave potassium cyanide. He reasoned that thunderstorms may have done the same thing. Several other reactions were listed showing how organic compounds could be produced from inorganic substances: carbon disulfide heated with hydrogen sulfide gave ethylene; and, as Berthelot had found, methane, carbon, and hydrogen reacted under an electric discharge to give acetylene, which could be oxidized to oxalic acid. Carbon monoxide heated with potassium hydroxide gave potassium formate; and from the distillation of formate Pflüger obtained methane, ethylene, butylene, amylene, and higher analogs. When ammonium formate was heated, hydrogen cyanide was formed. And ammonia passed over glowing carbon gave ammonium cyanide. Cyanides were known to be reactive compounds, so Pflüger postulated that cyanides could lead to proteins. He visualized that the origin of life was preceded by an intermediary period when lifeless organic compounds polymerized to become proteins of living nature.

Unfortunately, Pflüger lived at a time when the primary structure of proteins was still speculative. He proposed a theory for protoplasm in which the nitrogen was thought to be largely in the form of cyanogen. When Pflüger's cyan-protein theory failed to survive, so did his theory on the origin of life. Nearly fifty years passed before another significant proposal on the origin of life came to light.

In 1922, the Russian biochemist A. I. Oparin submitted a theory to the Botanical Society of Moscow on the self-assembly of abiogenetically-formed organic compounds as precursors of the first cells. These precursors were theorized to have developed by feeding on the abiotic organic compounds until they evolved to the point of being able to synthesize their own biochemicals. Oparin suggested that hydrocarbons were the initial organic compounds formed by the interaction of water and metal carbides found in the iron core of the earth. When the carbides and steam met at the surface of the earth the hydrocarbons resulted. He felt that they would have been burnt by the oxygen of the air, but the reaction would not have been complete, so that carbon monoxide and oxygen derivatives of hydrocarbons, the alcohols, aldehydes, ketones, and carboxylic acids, would have been formed. The nitrogen compounds, he believed, were derived from cyanogen formed by metal nitrites reacting with the hydrocarbons. The theory was published in 1924[2] and was followed by a revised edition in 1936.[3]

Oparin was influenced by the views of the time that held that the earth had been molten in its early stage and that the atmosphere remained essentially unchanged throughout history. When the British geneticist J. B. S. Haldane[4] gave support to the concept of chemical evolution in 1928, he expressed the idea of an atmosphere devoid of oxygen. To Haldane the agency of ultraviolet light acting on water, carbon dioxide, and ammonia created a variety of organic substances which accumulated in primitive oceans until there was the consistency of a "hot dilute soup."

Haldane's realization that it was an atmosphere without oxygen that was the key to the abiotic synthesis of organic compounds was a show of outstanding insight into the origin of life. But a curious flaw of human nature is to permit the imagery of a catchy phrase to shape one's reasoning. Haldane's hot dilute soup became "primordial soup," a feature that has been popularized for over fifty years without geologic evidence that it ever existed.

That life originated in a reduced environment was not generally considered before Haldane. It had been the consensus that the oxygenated conditions of today were necessary for life and that it was under these conditions that life began. But Pasteur, while studying fermentation, discovered that life without oxygen was possible and showed a host of anaerobic organisms that metabolized substances without it. In 1928, seeing that fermentative metabolic processes of living organisms are all similar, whereas the oxidative reactions are often quite different, Haldane concluded that anaerobic metabolism is the more primitive. He postulated that the first living organisms were anaerobes whose source of energy for metabolism was metastable molecules produced by the ultraviolet light from the sun.

Not until 1929 was it discovered that the most abundant element in the universe is hydrogen.[5] In a universe predominantly hydrogen, it was apparent that the earth's oxygenated atmosphere could not have been the original gaseous envelope. There were two ways, therefore, that the free oxygen could have been produced: by photodissociation of water in the upper atmosphere or by photosynthesis. The significance of the oxygen-generating photosynthetic organisms in the evolution of the earth's environment began to be realized. Oparin in his revised edition included the anaerobic origin of life—and it is now generally accepted that the development of life on earth created the oxidized state of the atmosphere and oceans.

Following the end of the Second World War the idea that the origin of life could be explained by the chemistry of the primordial earth began to draw adherents. The speculations were that life originated in some prebiotic organic milieu; the crux of the problem, however, was to find a plausible explanation of how organic substances could be produced from inorganic gases, salts, and minerals before any biological systems existed. Since plants construct compounds by reducing carbon dioxide, one of the approaches was to attempt to accomplish this by some physical means. But even if small compounds were obtained, they still had to be condensed into long chains before a functional cell could develop. J. D. Bernal,[6] a crystallographer and physicist at the University of London, wrote

The Physical Basis of Life in 1951 in which he suggested that absorption to clay surfaces could have been a means to bring about the concentration and polymerization of simple prebiotic compounds.

At this time the analytical technology had not yet reached a level that allowed the unraveling of the complex chemistry of the cell. Nucleic acids were known, but their exact significance was still uncertain; and proteins were considered to be the primary substance of life. The ultracentrifuge enabled biochemists to establish that proteins had definite weights, but whether they could ever be regarded as uniform compounds like giant chemicals was looked upon as doubtful—they had properties like colloids.[7] The first complete primary structure of a protein was not to be reported until Sanger[8] worked out the sequence of insulin in 1953—an achievement many biochemists up to that time would have considered impossible.

Although the concept of life originating by a combination of chemical processes on early earth appealed to scientific logic, there was no experimental data to substantiate how this could have been possible. At a time when the biosynthesis of proteins was still explained in terms of some vague and hypothetical template, an understanding of how something as complicated as a living cell could have arisen from inorganic minerals on primordial earth seemed remote indeed.

In 1951 a group at Berkeley led by Melvin Calvin[9] attempted to reduce carbon dioxide by using the cyclotron, the instrument developed for accelerating atomic particles to study the nucleus of the atom. It was an experiment to simulate the effects of radioactivity from the earth's crust on constituents of primordial earth to demonstrate an abiotic origin of organic substances. When a mixture of carbon dioxide in water with a catalytic amount of ferrous salt was bombarded with 40-meV helium ions from the 60-inch cyclotron, the chemists obtained a 4 percent conversion of carbon dioxide to formic acid and a 0.1 percent to formaldehyde. The experiment showed that some reduction had occurred. The results, however, were so unconvincing that the gulf between inorganic earth and proteins remained depressingly wide.

Two years later, in May 1953, Stanley Miller[10] published the results of his electric discharge experiment. Within a year Frederick Sanger announced the primary structure of insulin, the first protein ever whose amino acid sequence was determined. And about this same time James Watson and Francis Crick[11] showed that the chemical basis of life's reproductive nature was the double-helical nucleic acid.

Miller followed his first experiment with variations on the composition of the gas mixture and with detailed analysis of the material produced by the reaction.[12] A substantial amount of the product was nondescript, tarry material. Glycine was the amino acid produced in the highest yield—2.1 percent based on carbon. Of the long list of products generated by the reaction, only four were amino acids that occur in proteins.

Miller studied the mechanisms of the electric discharge experiment and dis-

Table 17.1. Yields from sparking a mixture of CH_4, NH_4, H_2O and H_2.

Compound	Yield (pM)	Yield (%)
Glycine	630	2.1
Glycolic acid	560	1.9
Sarcosine	50	0.25
Alanine	340	1.7
Lactic acid	310	1.6
N-Methyl alanine	10	0.07
α-Amino-n-butyric acid	50	0.34
α-Aminoisobutyric acid	1	0.007
α-Hydroxybutyric acid	50	0.37
β-Alanine	150	0.76
Succinic acid	40	0.27
Aspartic acid	4	0.024
Glutamic acid	6	0.051
Iminodiacetic	55	0.37
Iminoaceticpropionic acid	15	0.13
Formic acid	2,330	4.0
Acetic acid	150	0.51
Propionic acid	130	0.66
Urea	20	0.034
N-Methyl urea	15	0.051

Note: 59 moles (710) of carbon were added as CH_4. The percent yields are based on carbon.

Source: Stanley L. Miller and Leslie E. Orgel, *The Origins of Life on the Earth* (Englewood Cliffs: Prentice-Hall, 1974), p. 85.

covered that hydrogen cyanide (HCN) and aldehydes were the active interme-diates formed by the reaction. From this he deduced that the products were made by the Strecker reaction, the oldest known method of synthesizing amino acids. In order to test this hypothesis a mixture of HCN, aldehydes, hydrogen, and ammonia was boiled in water for one week. Products similar to those formed by the sparking experiment were obtained in respectable yields, supporting his ideas of the mechanism.

There would have been a number of energy sources on primordial earth that could have contributed to the synthesis of organic compounds. Among the most significant were ultraviolet light from the sun, electric discharges in the atmo-sphere, radioactivity in the soil, and thermal energy. From these, ultraviolet radiation in sunlight would have been in overwhelming predominance.

Today the ozone layer screens out virtually all radiation below 315 nanometers, but on primordial earth the solar flux would have streamed down with fierce intensity to have profound effects on the primitive atmosphere. Since the ab-sorption of methane extends to 145 nanometers, water to 165 nanometers, and ammonia to 220 nanometers, the photodissociation of these molecules would

Table 17.2. Boiling of H_2, NH_3, HCN, HCHO, CH_3CHO, C_2H_5CHO and H_2O for one week.

Reaction product	Yield Moles $\times$ 10^5	Percent
Glycine	98	16
Alanine	129	35
α-Aminobutyric acid	27	23
Glycolic acid	72	12
Lactic acid	42	12
α-Hydroxybutyric acid	15	13
Iminodiacetic acid	56	18
Iminoaceticpropionic acid	45	

have taken place at these shortwave ultraviolet emissions, creating highly reactive chemical radicals. Unfortunately, there are technical difficulties in using short-wave ultraviolet light in the laboratory to simulate the synthesis of organic compounds in a primitive atmosphere, and only some experiments at available wavelengths have been attempted.[13]

Philip Abelson[14] at the Carnegie Institute of Washington was the first to determine whether amino acids could be synthesized by photochemical reactions. When he radiated a solution of ammonium formate, ammonium hydroxide, sodium cyanide, and ferrous sulfate with the 254-nanometer wavelength, he obtained glycinonitrile, which hydrolyzed to glycine. Experiments were carried out by others using gaseous mixtures and several ultraviolet wavelengths in which

Table 17.3. Present sources of energy averaged over the earth.

Source	Energy (Calories/cm²/year)
Total radiation from the sun	260,000
Ultraviolet light	
<300 nm	3,400
<250 nm	563
<200 nm	41
<150 nm	1.7
Electric discharges	4
Cosmic rays	0.0015
Radioactivity (to 1.0 km depth)	0.8
Volcanoes	0.13
Shock waves	1.1
Solar wind	0.2

Source: S. L. Miller, H. C. Urey, and J. Oró, Origin of organic compounds on the primitive earth and in meteorites, J. Mol. Evol. *9*, 59–72 (1976).

other amino acids were formed.[15] Glycine, alanine, valine, and leucine were identified as products from irradiated methane, ammonia, water, and carbon monoxide in the region between 145 and 180 nanometers.[16] The amino acids formed by ultraviolet light were generally those found from the electric discharge experiments.

Although ultraviolet light may have been a predominant source of energy for the synthesis of prebiological organic compounds, most of it is of long wavelengths that do not react with the gases used to simulate the primordial atmosphere. Carl Sagan and Bishun Khare, however, observed that hydrogen, methane, ammonia, water, ethane, and perhaps carbon monoxide and nitrogen are all entirely transparent to ultraviolet light above 240 nanometers, but that hydrogen sulfide has a broad absorption continuum beginning at about 270 nanometers. Hydrogen sulfide then could have served as the initial photon acceptor to capture energy of the long wavelength ultraviolet light and initiate chemical reactions. In simulation experiments, Sagan and Khare[17] found the presence of hydrogen sulfide and ethane with the gases irradiated by ultraviolet light of 254 nanometers did produce amino acids, thus showing that the abundance of ultraviolet light in the longer wavelength region of the spectrum may have been one of the principal sources of energy for the synthesis of prebiotic compounds.

Ionizing radiation can also serve as an energy source. When a mixture of gases simulating the primitive atmosphere was exposed to X-rays and gamma rays, amino acids were formed.[18] This energy source is not viewed as having been a significant contributor to the synthesis of prebiotic compounds, however, since most of the ionization would have taken place in silicate rocks and not the atmosphere.

The importance of volcanic activity in the synthesis of amino acids has been disputed. When a mixture of gases simulating a primordial atmosphere is heated continuously at 1,200°C, no organic compounds are formed and only an equilibrium of the gases results. If, however, the gases are passed over molten lava slowly and then are rapidly quenched, a variety of hydrocarbons are synthesized. In this way phenylacetylene, a precursor to phenylalanine and tyrosine,[19] and indole, the precursor of tryptophan,[20] are formed. This reaction presents a manner in which the aromatic amino acids could have occurred prebiologically. But these amino acids may not have become a part of biological systems until they were synthesized biochemically.

As curious as it may seem, it is conceivable that a principal producer of amino acids was shock waves. A. Bar-Nun and colleagues[21] have demonstrated with gases simulating a primitive atmosphere that in a shock tube the shock front temperature rises to between 1,000 and 2,000°K within 10^{-8} second with rapid quenching, giving a high conversion of the gases to amino acids. The authors calculated the annual shock flux from cometary meteors, micrometeorites, and thunder to be approximately one calorie per square centimeter. In contrast, the energy form ultraviolet light below 300 nanometers would have been about 1,000 calories per square centimeter. But shock waves are a million times more efficient in producing amino acids than ultraviolet light. In this context, there would have been 1,000 times more amino acids generated by shock waves than by the most abundant energy source, the ultraviolet light.

These various experiments suggested a scenario in which organic compounds or their precursors were produced in the primordial atmosphere and rained down to accumulate in the juvenile oceans. With many reactive compounds washed into catchbasins, interactions would have generated additional substances. It was an exciting new image of the chemistry that took place on earth before life existed. The origin of life was no longer thought of in terms of strange inorganic substances in a geologic setting giving rise to a non-biological structure that had to undergo an undetermined evolution before a living cell could be formed; the prebiotic compounds needed to construct a biological system were the amino acids familiar to every biochemistry student.

After three decades it is possible to put the Miller experiment in a better perspective in respect to the origin of life. Urey's methane–ammonia atmosphere was calculated as the primary atmosphere of the earth, acquired when the solar system was formed. But there is ample evidence to indicate that the original atmosphere was lost and replaced by outgassing of the earth's interior. The question then remains: In what manner did the second atmosphere differ from the first?

Studies have shown the composition of volcanic gases to be principally water and carbon dioxide with a significant amount of hydrogen.[22] Chloride, as the substance third-most in abundance, issues as the acid, as does sulfur; nitrogen is present principally as the element. Ammonia could not have been in the atmosphere in any significant quantity. Abelson[23] pointed out that a quantity of ammonia equivalent to the present atmospheric nitrogen would have been degraded by ultraviolet radiation in approximately 30,000 years. Similarly, its high solubility in water would have created an equilibrium with essentially all of it in solution in the oceans. The early earth's atmosphere, therefore, consisted of residual volcanic gases after they equilibrated between the atmosphere–ocean system. Abelson contends that its composition would have been principally carbon monoxide, carbon dioxide, nitrogen, and hydrogen.

The fundamental nature of the Miller experiment allows the atmosphere of precursor gases to be apparently of any composition provided it is reducing and the components contain C, H, O, and N. Nevertheless, the chemistry for the outgassed atmosphere may have been similar to that for the methane–ammonia mixture but not the same. When a mixture of carbon monoxide, nitrogen, and hydrogen is subjected to an electric discharge, hydrogen cyanide (HCN) is the major product, with little else except carbon suboxide ($O=C=C=C=O$), carbon dioxide, and water.[24] One does not obtain the series of aldehydes that was

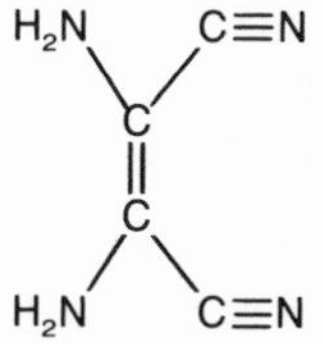

$$H_2N \quad C\equiv N$$
$$\diagdown \quad \diagup$$
$$C$$
$$\|$$
$$C$$
$$\diagup \quad \diagdown$$
$$H_2N \quad C\equiv N$$

Figure 17.1 Diaminoma-
leonitrile (HCN tetramer).

intermediate to the amino acids by the Strecker reaction. As a result, the prebiological chemistry becomes one not of a "soup," as envisioned resulting from
a methane–ammonia atmosphere, but of HCN.

Hydrogen cyanide undergoes two important reactions: it hydrolyzes to ammonium hydroxide and formic acid; and in slightly alkaline solutions it condenses
with itself to yield a variety of compounds of biological interest. The major
product from the condensation of HCN is the tetramer. Hydrolysis of the tetramer
yields glycine, and J. Oró and S. Kamat[25] reported small amounts of alanine
and aspartic acid when 2.2 molar cyanide solutions were heated at 70°C for 25
days. But this is a high concentration. And in order for the tetramer formation
to exceed the hydrolysis reaction, the concentration of hydrogen cyanide needs
to be greater than around 0.1 molar. This is an unrealistic concentration under
natural conditions for hydrogen cyanide, but it could have been commonplace
for ammonium hydroxide from hydrogen cyanide hydrolysis to combine with
additional hydrogen cyanide to give ammonium cyanide, which could then be
concentrated by partial evaporation of the solution.

It is possible that this concentration factor may have been circumvented by
the action of ultraviolet light. Abelson found that 0.002–0.10 molar solutions
of hydrogen cyanide at pH 8–9 irradiated with 254-nanometer light gave a product
mixture that hydrolyzed to glycine, serine, aspartic acid, and glutamic acid.

The research by the various laboratories in the wake of Miller's experiment
was actively pursued in an effort to show the abiotic synthesis of as many of
the 20 biological amino acids as possible. As a result, all but histidine and
arginine have been reported to have been produced in simulation experiments.[26]
In addition to these, the Miller experiment yields many amino acids that are not
used by proteins.

Life may have begun with much less. In one respect the prebiological chemistry
based on hydrogen cyanide is simpler and perhaps more coincidental with the
origin of life. When we look at the process of contemporary biosynthesis, we
are impressed by how few basic building blocks there really are.[27] Pyruvate,
acetate, and carbonate in some microorganisms can furnish all the carbon for:

serine	leucine	methionine
glycine	isoleucine	glutamic acid

cysteine	lysine	proline
alanine	aspartic acid	arginine
valine	threonine	

Pyruvate and acetate can result from the degradation of serine; and from pyruvate and from malonic acid, acetate can be obtained. Furthermore, ferredoxin, which is believed to have been one of the earliest proteins, was apparently formed from a repeating sequence of alanine, serine, aspartic acid, and glycine—amino acids that are produced most readily from hydrogen cyanide.

The Miller experiment illustrated the ease of formation and the probability of the prebiotic presence of life's building blocks; and a number of energy sources were available on primordial earth for the production of biological compounds. A requisite was a mixture of reduced gases. Ironically, if the atmosphere had originally contained oxygen, which had long been believed necessary for the origin of life, life never would have happened. Oxygen would have quenched the reaction before it could have made the building blocks.

Abelson pointed out that an atmosphere containing appreciable amounts of methane would have resulted in the production of a large deposit of hydrophobic organic material, but that unusually large amounts of carbon or organic material have not been found in rocks predating the earliest organisms. There are indications, however, of a considerable number of organic compounds that apparently were synthesized in a methane–ammonia atmosphere—not on primordial earth but in the gaseous nebula before the earth's formation.

Hydrocarbon gases have been found near Lake Huron in Canada and as the Ukhtinsk deposit in the USSR in crystalline rock formations having no connections with sedimentary rocks.[28] The great reservoirs of petroleum are definitely of biological origin, but findings in the Soviet Union clearly indicate that some petroleum deposits are of an abiotic derivation. Gas emissions in the Khibinsk massif on the Kola Peninsula show methane and heavier hydrocarbons that apparently are of magmatic origin. Although they number in the hundreds, such sources rarely have commercial importance.[29] It appears that this organic matter is not a vestige of the "primordial soup," but has survived from a time even before the earth existed. This hypothesis stems from and is supported by the discovery of organic compounds in meteorites.

18

Nucleosides, Nucleotides, and ATP

When Miller reported his electric discharge experiment in 1953, protein chemistry was the main thrust of biochemistry, and molecular biology was still a young branch. As a result of this interest in proteins and the relative ease with which minute amounts of amino acids could be measured, the initial emphasis in origin of life studies was on amino acids. But proteins are not, nor could ever have been, life unto themselves. Biological reproduction fundamentally is molecular replication, a role that is performed exclusively by nucleic acids. Where then were the nucleotides—the building blocks of nucleic acids?

Nucleotides are composed of three types of chemicals: a heterocyclic base, a sugar, and phosphoric acid. The bases are the purines—adenine and guanine; and the pyrimidines—cytosine, uracil, and thymidine. The sugar is either ribose or deoxyribose. Before considering the formation of DNA's building blocks, we must account for the prebiotic presence of these three structural components.

The Miller experiment has shown the bridge from the gaseous constituents of a primordial atmosphere to the organic substances of living systems. Fixation of carbon dioxide and nitrogen apparently had been a common occurrence through the action of natural sources of energy by a reaction in which the intermediate product was hydrogen cyanide. If, instead of concentrating our attention on amino acids, we consider the potential role of hydrogen cyanide, the scenario takes on a wider scope.

Hydrogen cyanide hydrolyzes to ammonium hydroxide and formic acid, but additional hydrogen cyanide with ammonium hydroxide leads to ammonium cyanide. The wash of primordial skies into shallow lakes and basins with periodic evaporations to or near dryness could have led to large amounts of this chemical having accumulated and concentrated on a youthful earth.

In 1960, Juan Oró[1] at the University of Houston heated a concentrated solution of ammonium cyanide for one day at 90°C. When he later analyzed the sample for products of possible biological significance, he discovered that he had synthesized adenine, the principal heterocyclic base in nucleic acids and many important coenzymes.

It appeared that 4-aminoimidazole (AICA) and formamidine were the probable intermediates. In a follow-up experiment Oró and Kimball[2] heated a solution of AICA to temperatures between 100 and 140°C and obtained two other purines, guanine and xanthine, each in a 1.5 percent yield. In another experiment, with S. S. Kamat,[3] hydrogen cyanide added to 3 Normal ammonium hydroxide and heated at 70°C for 25 days yielded glycine, alanine, and aspartic acid among the products. This synthesis of adenine and amino acids from ammonium cyanide was confirmed by C. U. Lowe and others,[4] who reported as many as 75 ninhydrin-positive products from the reaction (see next page).

On primordial earth some purines may have been forming even before evaporation concentrated the ammonium cyanide. Leslie Orgel and his group[5] at the Salk Institute in La Jolla, California, have shown that ultraviolet light transforms the hydrogen cyanide tetramer to monoaminomalonitrile. This important intermediate can react with additional hydrogen cyanide to yield adenine or it can hydrolyze to AICA, which can condense with cyanide, formamidine, or cyanogen to produce guanine and xanthine.

The role of hydrogen cyanide took on additional significance in 1977, when J. P. Ferris, J. C. Joshi, and J. G. Lawless[6] isolated pyrimidines from the hydrolysis of a cyanide experiment. A 0.1 molar solution of hydrogen cyanide was adjusted to pH 9.2 with ammonium hydroxide and allowed to stand at room temperature for 4–12 months. When the reaction mixture was then hydrolyzed and analyzed, yields of 4,5-dihydroxypyrimidine and 5-hydroxyuracil were obtained.

The image of ammonium cyanide and oligomers of hydrogen cyanide as precursors of purines, pyrimidines, and amino acids gave a satisfying unity to prebiological chemistry. Instead of the building blocks of proteins and nucleic acids being formed separately and brought together through circumstances, both formed *in situ* from the same substances.

Another synthesis of pyrimidines in a simulation experiment was achieved by the group at the Salk Institute[7] using cyanoacetylene, the second most prevalent intermediate produced by the electric discharge experiments. Sanchez and others reported that a 0.1 molar concentration of cyanoacetylene in a 1 molar solution of potassium cyanide heated at 100°C for one day gave a 5 percent yield of cytosine; the reaction at room temperature for seven days yielded 1 percent. Once cytosine was formed, uracil followed by an easy hydrolysis (see reaction equation at top of page 162).

The formation of hydrogen cyanide and cyanoacetylene from the gases of primitive earth could demonstrably have given ammonium cyanide, the hydrogen oligomers, and ultimately the amino acids, purines, and pyrimidines necessary for the biologically important building blocks to proteins and nucleic acids. The heterocyclic bases, however, have to be condensed with a sugar and phosphoric acid to make the nucleotides for the polymerization to nucleic acids.

At first glance, the prebiotic synthesis of sugars would appear to be the easiest of all simulation experiments to devise. Formaldehyde is one of the common substances formed in the prebiotic experiments, and as early as 1861, Buterlow had shown that formaldehyde undergoes condensation in alkaline solutions to

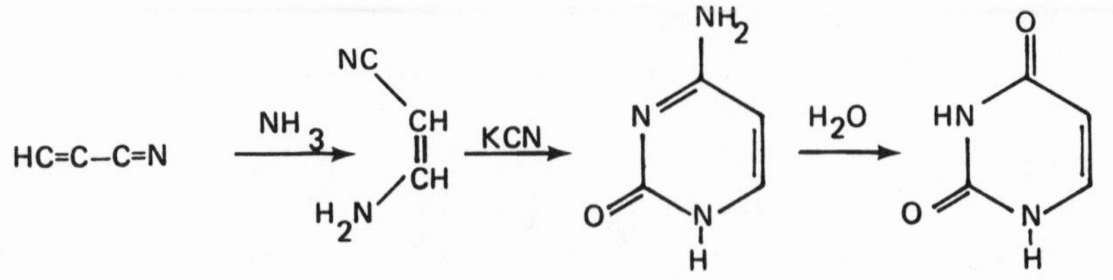

Cyanoacetylene   Cytosine   Uracil

Monoaminomalonitrile

produce sugars. After the formation of glycoaldehyde the synthesis proceeds to the higher sugars, the tetroses, pentoses, and hexoses, including ribose and glucose.

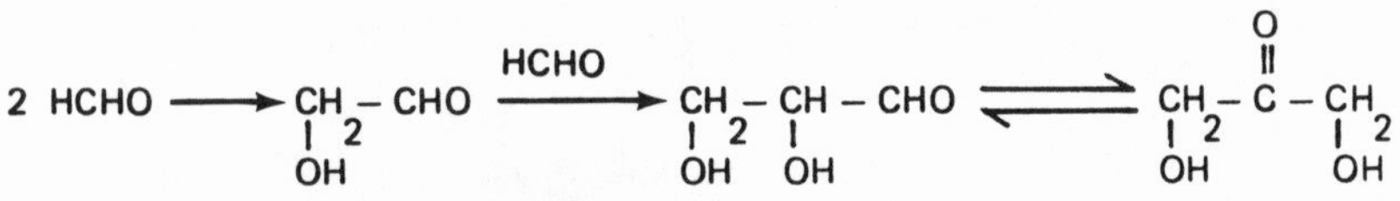

Formaldehyde Glycoaldehyde

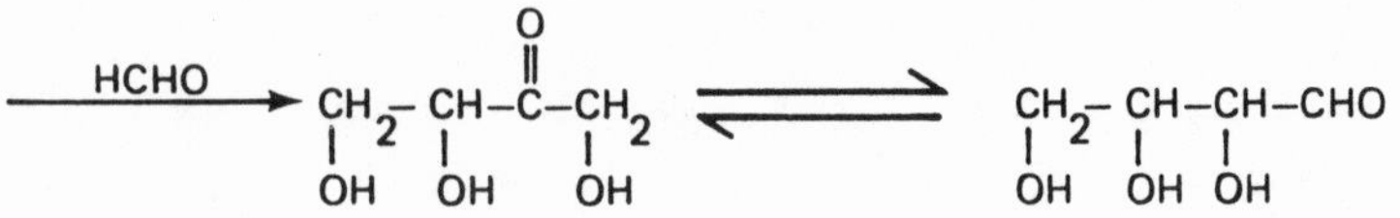

As attractive as the condensation of formaldehyde seems for the prebiotic synthesis of sugars, there are serious chemical objections. The reaction occurs only in formaldehyde concentrations greater than 0.01 molar; and monomeric sugars are unstable in water, especially if the pH is much below 7, decomposing to lactic acid as the major product. A further problem is that sugars react with amino acids by the browning (Maillard) reaction to form nonbiological products. Since it has to be assumed that the simplest arrangement for the origin of life would be where all the starting components were formed and present together at the same time, interactions that would generate inactive products have to be taken into consideration. This would appear to be enough to exclude sugars, but actually only two sugars are vital for the initial assembling of a functional cell. They are ribose and deoxyribose, the essential components of the nucleotides. In this capacity, the chemical problems with sugars disappear; ribose bound in the form of nucleotides no longer has the intrinsic instability of sugars.

    J. Oró and A. C. Cox[8] investigated relevant conditions where ribose and deoxyribose could have been formed on primitive earth. They learned that aqueous solutions of acetaldehyde (CH_3CHO) with formaldehyde (HCHO) or glyceraldehyde ($CH_2OH \cdot CHOH \cdot CHO$) and catalyzed with calcium oxide at 50°C gave a 3 percent yield of deoxyribose. Ammonium hydroxide proved to be an even more satisfactory catalyst.

The generation of formaldehyde and acetaldehyde by reactions on a primordial atmosphere would have required the presence of hydrocarbon gases. Since the occurrence of hydrocarbons in some igneous rocks and meteorites suggests a prebiological presence of methane and its homologs, there is reason to believe that the precursors to the aldehydes and fatty acids were present.

Coupling of adenine and deoxyribose to form deoxyadenosine occurs readily. When the two reactants were allowed to stand together at room temperature in the presence of cyanide, a 1 percent yield of deoxyadenosine was formed;[9] with phosphate present the yield increased to 5 percent; when ultraviolet light was used with cyanide, 7 percent was obtained. Unfortunately, the other bases did not produce their nucleotides by this reaction, but these could have been formed by other means. A mixture of seasalts, for instance, is an effective catalyst for forming nucleosides when purines are heated with ribose, giving yields on the order of 1.5 to 4.5 percent.[10]

This procedure does not work for pyrimidines. Although pyrimidine nucleosides do not form from the direct combination of the base and ribose, they conceivably could have arisen less directly. When ribose-5-phosphate was equilibrated with ammonia to form ribosylamine, and then treated successively with

α -5'-Cytidylic acid $\qquad\qquad$ β -5'-Cytidylic acid

cyanogen and cyanoethylene, the main product was α-cytidylic acid. The biological form, however, is the β isomer. Sunlight, which has been implicated in most of the simulation experiments, plays a decisive role here. When α-cytidylic acid was radiated with ultraviolet light, as much as 10 percent was converted to the β-cytidylic acid, the isomer used by biological systems.[11]

To proceed from the nucleoside to nucleotide involves phosphorylation, or attachment of a phosphoric acid group to the sugar moiety. This step is closely connected with the general problem of the prebiological formation of phosphate derivatives. The compound that normally transfers chemical energy in biological systems is ATP. And it is the energy from the hydrolysis of the pyrophosphate bond degrading ATP to ADP that fuels the reaction of most biochemical syntheses. Pyrophosphates were essential from the beginning, and ATP has the chemical stability and reactivity to make it most appropriate for the task.

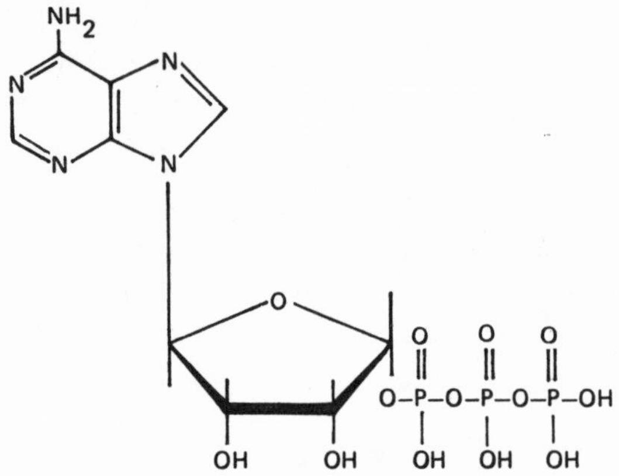

Figure 18.1. The structure of adenosine triphosphate (ATP).

Since nucleic acids are polymers of nucleotides joined through the phosphate group in a 3'-5'-linkage of their sugar groups, a prebiological means of phosphorylating nucleosides must have existed on primordial earth. If one can justify the geologic occurrence of phosphoric acid and its derivatives on primitive earth, the simulation experiment is not difficult. Dissolving nucleosides in polyphosphoric acid at 0–22°C gives 25 to 45 percent yields of the monophosphates.[12] But the possible occurrence of polyphosphoric acid under geologic conditions is too remote to be considered as a plausible answer.

The difficulty with phosphorus is the extreme insolubility of its calcium salt, apatite. Since calcium exceeds the abundance of phosphorus in basalt by approximately 20-fold (10-fold in granite), phosphorus is effectively removed from seawater by precipitation of the calcium salt. On the other hand, phosphorus is not a rare element and makes up as much as 0.6 percent of the mineral content of igneous rocks.[13] And although hydroxyapatite may form in the precipitation of phosphorus from pure aqueous solutions, in a marine environment phosphorus is deposited as a carbonate fluorapatite. As a result, the concentration of phosphorus in seawater remains extremely low at 2.2 micromolar.

The problem then is to find circumstances compatible with primitive earth in which phosphorus could have been available to be incorporated into organic compounds: in other words, to demonstrate a chemical form of the element which could have been capable of phosphorylation reactions. Stanley Miller and M. Parris[14] succeeded in producing pyrophosphate on the surface of hydroxyapatite with cyanate salts. Since calcium pyrophosphate is as insoluble as apatite, any synthesis would have had to have been on the crystal surface. This is conceivable but not really convincing; a more soluble form of phosphate with the emergence of the functional cell would appear more favorable.

Alan Schwartz[15] of the University of Nijmegen, The Netherlands, proposed a procedure in which the phosphate could have been solubilized and concentrated within a functional level. Oxalic acid is a strong complexing agent for calcium. If oxalic acid, which could have been formed by the decomposition of glycine or formic acid or by the hydrolysis of cyanogen, reaches a concentration of 0.01 molar at pH 5 in contact with apatite, then it frees phosphate to the extent of 0.003 molar, a relatively large amount of phosphate in solution. Schwartz feels that cyanogen, being a prebiological water-soluble gas, could have provided a constant supply of oxalate at every rock surface exposed to weathering and that phosphoric acid derivatives would have been available by this means.

Once the difficulty of accounting for the geologic presence of phosphoric acid salts is overcome, phosphorylation reactions are known that could have occurred. When a nucleoside is heated from 50°C to 160°C with a number of hydrogen phosphate compounds, the mononucleotide results.[16] But a more effective phosphorylation occurs in the presence of urea. Nucleosides heated to 100°C in a dry mixture of urea, sodium hydrogen phosphate, ammonium chloride, and ammonium bicarbonate are converted to phosphorylated derivatives in yields in excess of 90 percent.[17]

Urea has been a product of most simulation experiments of prebiotic conditions and must have been common on primitive earth. It would have been present in any exposed solutions of ammonium cyanide since ultraviolet radiation promotes the conversion of ammonium cyanide to urea. And on further studies of urea-catalyzed phosphorylation reactions, L. Österberg, L. E. Orgel, and R. Lohrman[18] discovered that when phosphate exceeds nucleosides in the mixture, pyrophosphate bonds are formed. Four days of heating at 100°C gave a 15 percent yield of thymidine triphosphate.

It appears then that the nucleoside triphosphates arose prebiologically under plausible geologic conditions. The essential requisite was the occurrence of the phosphate as its acid salt. To visualize how this could have come about we need to return to the setting of the earliest microorganisms on earth.

The Onverwacht Series rocks in which the microfossils occur are cherts that lie on top of extrusive, subaqueous volcanic rocks. Although the cherts are found in horizontal extensions, more often they occur as pockets in the surface of the ancient lava. What occurred then is probably the same as what happens today with many such extrusives. The juvenile water brought to the surface either at the time of eruption or in fumaroles is extremely rich in nutrients—it is charged with carbon dioxide and the essential minerals of phosphorus, sulfur, and nitrogen. Presumably the high temperature of volcanism liberates the phosphorus from the minerals in the igneous rocks and brings it to the surface as soluble acid salts, where it enriches the waters of volcanic lakes.

The same phenomenon is evident today where the microbiota of volcanic pools often grow so profusely that they form crusts which eventually cover the entire surface of these pools. When the lakes are large like Waimungu in New Zealand, the growth creates crusts at the edges and extends for great distances toward the center of the lake.[19] It is quite possible that life began on earth over 3.5 billion years ago not in an ocean with the consistency of a "primordial soup" but in the phosphorus- and sulfur-rich volcanic pools that pockmarked the earth's surface during the Archean.

19
Polypeptides

The discovery of the manner in which amino acids and nucleotides would have been formed on primordial earth bridged the gap between the inorganic geology and the basic organic units of biological systems. But having crossed one bridge, we find ourselves facing another chasm before we can reconstruct life's chemical origins. There is no quality of life in amino acids and nucleotides. A living system comes not from the building blocks, but from coordinated interaction of their biopolymers; for it is the relationship between nucleic acids and enzymes that constitutes the foundation of biology.

Therein lies the problem. In organisms as we know them, all biochemical reactions are catalyzed by enzymes, including those reactions that lead to the synthesis of the nucleic acids and enzymes themselves. If enzymes are needed to create the nucleic acids and enzymes, both of which are needed to make enzymes, where then did the initial enzymes come from?

Just as the attitude of the 1950s placed an emphasis on amino acids as the most significant building blocks, so too it seems there was an emphasis to find an abiotic origin of proteins, and consequently enzymes, on the belief that once the enzymes were formed, the synthesis of all the other components was possible. Perhaps it was more conceivable to think of the abiotic appearance of a polypeptide with enzymatic properties than to imagine a gene, which can occur in contemporary organisms with a molecular weight in the millions, to have formed outside a biological system. But it seems likely that the attention paid to proteins was due to the fact that biochemists were more familiar with amino acids and the established chemistry of polymerizing them to polypeptides in the laboratory.

Even before F. Hofmeister[1] and E. Fischer[2] confirmed the polyamide nature of proteins, chemists were trying to polymerize amino acids by heating them together. In 1897, Hugo Schiff[3] condensed aspartic acid at temperatures of 190–200°C; and three years later, L. Balbiano and D. Trasciatti[4] prepared polyglycine by heating the amino acid in glycerol. When heated without a solvent, glycine gave a mixture of peptides and glycine anhydrides.[5] Certainly the most successful application of thermal polymerization of amino acids was by Wallace Carothers. In 1936, while working for the DuPont Company, Carothers heated omega-amino acids, where the amino group is on the carbon atom farthest from the caraboxyl end, and obtained a polyamide polymer which became known as nylon.[6]

The polymerization reaction is a dehydration step, the reverse of hydrolysis. That is to say, a molecule of water has to be removed for each peptide linkage formed when condensing the amino acids. A logical approach to achieving polymerization, therefore, would be to find applicable dehydrating conditions. The oft-cited proposition that macromolecules were formed from the condensation of the monomers prebiologically in a primordial ocean is naive because it doesn't take into account the thermodynamic requirement of 8–16 kilojoules per mole to form the peptide bond. This thermodynamic barrier is too high for the monomers to merely condense by themselves in an aqueous environment. The fact that organisms grew and evolved in the oceans, however, does not necessarily mean that the essentials of the first functional cell were formed there.

One of the earliest attempts to find a prebiological origin of enzymes or enzymelike amino acid polymers was by Sidney Fox, presently at the University of Miami. In a paper written with M. Middlebrook[7] and presented to the American Society of Biological Chemists in 1954, he reported that heating certain amino acids at 200°C for one-half to three hours gave anhydropolymers. Fox compared the results of these condensations to an earlier paper of his describing a selectivity of amino acids to couple with each other in an enzyme-promoted condensation.[8]

Then in 1959, S. W. Fox, K. Harada, and A. Vegotsky[9] reported on the thermal polymerization of amino acids in a paper espousing a theory that the thermal synthesis of biochemicals preceded the origin of life in a prebiological evolution resembling the pathways of biosynthesis. For instance, when malic acid was heated with urea or ammonia, aspartic acid resulted.[10] The aspartic acid could be degraded in turn to α- or β-alanine. Furthermore, aspartic acid reacted with urea in the presence of calcium or magnesium hydroxide to give ureidosuccinic acid, an intermediate in the biosynthesis of pyrimidines. By heating the salt of an acid from the Krebs cycle, they were generating some products that formed biogenetically by a similar pathway. It was an attractive hypothesis, but the correlation applied only to a few simple structures and did not extend farther along the biosynthesis pathways.

Biochemists knew that when a mixture of amino acids in the ratios found in proteins was heated, the result was pyrolysis to a dark brown tar with a disagreeable odor. Since aspartic acid polymerizes readily by heating but glutamic acid does so only reluctantly, Fox and Harada discovered that glutamic acid mixed and heated with aspartic acid copolymerized with it. Except for glycine, the neutral amino acids, however, do not form homopolymers because of steric hindrance. But by using a mixture of amino acids that was predominantly aspartic and glutamic acids with an equimolar mixture of the remaining 16 amino acids in a 2:2:1 ratio, Fox and Harada were able to heat an amino acid mixture and obtain a nondialyzable product that analyzed for all of the amino acids to some extent.

The procedure was reported in 1958 as the thermal polymerization of amino acids to a product resembling protein.[11] An amino acid mixture containing 80 percent aspartic and glutamic acids heated in an oil bath at 170°C for three hours gave a glasslike product. After this material was dissolved in water, dialyzed, and lyophilized, the resulting brown powder gave an analysis for a mean chain weight (molecular weight) of 4,900 and contained 71 percent aspartic acid, 15 percent glutamic acid, and 14 percent of all of the other amino acids. From the N-terminal amino acid analysis, the conclusion drawn was that there was a nonrandom arrangement of the monomers in the polymer.

This notion that a polymer resembling protein could be made by heating a proper proportion of amino acids received considerable attention in the 1960s, and a large number of papers attempting to correlate many of the physical and chemical properties of proteins to these thermal polymers were published.[12] Most of the comparisons were of uncritical properties such as color tests or elemental analyses due to amino acids or short peptide linkages. On the other hand, these thermal polymers failed to elicit any immune response when tested for antigenicity,[13] an extremely sensitive test for protein structure.

Fox coined the name ''proteinoid'' for these thermal polymers, which he contended contained all of the 18 amino acids found in proteins (glutamine and asparagine are excluded). The types of proteinoids was extended to basic ones which were predominantly lysine, an amino acid that can also form a homopolymer. A series of extremely weak catalytic properties were associated with various thermal polymers and were suggested to be ''enzyme-like'' activities.[14]

Because of their heterogeneity, it was impossible to determine the amino acid arrangement in these thermal polymers by sequencing, as is performed routinely with natural peptides. As a result, much of Fox's interpretation regarding the structure was directed to show that there was ''limited heterogeneity'' and nonrandomness, the implication being that amino acids heated together condensed with a degree of selectivity to their order to form sequencing resembling that found in proteins. Nonrandomness and unequal proportions of amino acids in a thermal polymer, however, are no indication whatever that the arrangement of amino acids resembles proteins. There are simpler explanations that are consistent with chemical behavior.

Chemically, aspartic acid is the amino derivative of succinic acid.[15] It polymerizes readily upon heating because it can form an anhydride which then reacts with any available amino group, including that on other aspartic acids. Once the

polymerization is initiated, it proceeds until it exhausts the available aspartic acid
or is terminated by a neutral amino acid which cannot extend the process.

By using amino acid mixtures consisting predominantly of either aspartic acid
or lysine—amino acids that polymerize to large homopolymers easily—Fox
obtained polymerization of these amino acids with small amounts of the others
incorporated. The high molecular weights reported for proteinoids alleged to be
analogous to proteins were attained only if the polymer was essentially polyaspartic acid or polylysine. A feature of the reaction was that the larger the proportion of neutral amino acids, the lower the average molecular weight of the
product—a clear indication that these amino acids, unable to continue the polymerization, terminated it when they reacted.

At first glance the amino acid analysis of these thermal polymers looks like
the analysis of a standard protein with the mole-percent for all 18 of the amino
acids listed. The fallacy in the interpretation of the analysis, however, lay in the
fact that the molecular weights of these thermal polymers were only 4,000 to
5,000. With a molecular weight of 4,500, the analysis indicated that the polymer
contained 22 aspartic acids, 3 glutamic acids, and 8 of the others;[16] when the
molecular weight was 4,900, the calculated composition was 26 aspartic acids,
8 glutamic acids, and 7 others. In other words, despite giving an analysis for
all 18 amino acids, neither polymer was large enough to contain more than 7
or 8 of the other 16 amino acids in addition to the aspartic and glutamic acid in
one molecule. Many of the amino acids, such as tyrosine or phenylalanine,
analyzed in a percentage so low that the molecular weight of the thermal polymer
would have had to have been 3 or 4 times larger to contain even one residue.
It was clear that the results could not indicate copolymerization of 18 amino
acids in the same polymer, and by definition the thermal polymers were not even
proteinoids.

There can be little doubt that the biological synthesis of proteins cannot be
simulated by merely heating a heterogeneous mixture of the amino acids. Steric
hindrance prevents amino acids with hydrophobic side-chains from reacting in
thermal reactions, except at terminal positions. In effect, only those amino acids
known to polymerize easily, do so. Proteinoids were mixtures of polyaspartic
acid or polylysine interspersed with glutamic acids and glycines and with neutral
amino acids terminal. Extended heating or higher temperatures increased the

number of neutral amino acids incorporated, which was consistent with the imide bond of condensed aspartic acid reacting with these amino acids, adding them to the side-chain. The "copolymerization" of all 18 amino acids was based on the amino acid analysis of the product, which showed that all were present to some extent. But this indicates nothing about the sequence. When polyaspartic acid is heated with a similar mixture of amino acids, the resulting polymer also gives an analysis for all the amino acids in the same manner as the so-called proteinoids.[17]

The Miller experiment gave amino acids because the products formed by the electric discharge were washed into the reflux flask, where they hydrolyzed readily. This dropped the intermediate substances back behind a thermodynamic barrier to spontaneous polymerization and set up the necessity for an energy-demanding condensation reaction to produce a peptide. Rather than trying to invoke anhydrous locales or high temperatures to reinstate the condensed material, some researchers have investigated methods of obtaining polypeptides more directly.

The answer to the formation of peptides may lie in the Miller experiment itself. When the sparking experiment was carried out with the water kept at 40°C, peptides as well as amino acids were isolated from the solution.[18] When the experiment was conducted at room temperature, amino acids were detected only after hydrolysis of the solid residue recovered after evaporating the solution.[19] This temperature effect was also noticed by Sagan and Khare[20] in their photochemical studies, where solids hydrolyzable to amino acids were obtained below 50°C. These results seem to indicate clearly that amino acids are not formed directly in the reaction but are produced only through hydrolysis of their precursors.

The precursors of the amino acids, therefore, existed in some condensed form that gave amino acids on hydrolysis. Hydrogen cyanide polymerizes in basic solutions to yield a mixture that includes the tetramer (diaminomalonitrile), a pentamer (adenine), polymeric amino acid precursors, and black, intractable solids that are believed to have fused tetrahydropyridine structures. Clifford Matthews and Robert Moser[21] isolated peptidelike solids from the reaction mixture and hydrolyzed them to 12 of the 20 amino acids common to proteins. They later reported the formation of similar peptide products from the hydrolysis of the tetramer diaminomalonitrile.[22]

In subsequent work, Matthews,[23] now at the University of Illinois at Chicago Circle, proposed that heteropolypeptides are formed directly from hydrogen cyanide and would have been formed on primordial earth. According to the hypothesis, there exists a low energy pathway for hydrogen cyanide to polymerize readily to polyaminomalonitrile. Successive reactions of hydrogen cyanide with the reactive nitrile groups should then yield heteropolyamidines, which can react with water to produce heteropolypeptides with sequencing of amino acids similar to that of proteins.

There is spectroscopic evidence that peptide bonds and nitrile groups exist in products formed by ionizing radiation in aqueous cyanide solutions.[24] And subsequent hydrolysis of these products yielded several amino acids. In support of the model for the origin of proteins, Matthews and his group[25] have reacted poly-α-cyanoglycine with hydrogen cyanide and obtained a product that hydrolyzed to various amino acids, including glycine, alanine, valine, and aspartic and glutamic acids.

But the conclusion that heteropolypeptides were actually formed as the precursors has been contested.[26] Tests that are specific for the presence of peptide bonds, such as catalysis by pronase, an enzyme that can hydrolyze diglycine to glycine, have been negative. Whether heteropolypeptides are formed by this method and have sequencing like that of proteins probably will remain disputed unless a product is obtained that responds in a biological reaction, such as enzymic degradation. C. I. Simionescu, F. Dénes, and others[27] in Romania have been investigating the preparation of polymeric substances in simulation experiments under conditions of cold plasma. A mixture of methane, ammonia, and water vapor was introduced through a needle valve into a chamber, where the gases were exposed to an electric discharge. The products deposited on the wall and bottom of the cylindrical flask, which had been cooled to -60 to $-40°C$. Fractionation of these products yielded polymeric fractions with molecular weights in the tens of thousands. And when the raw material was hydrolyzed, amino acids, purines, pyrimidines, and a number of neutral and basic ninhydrin-positive derivatives of unknown structures were found in the hydrolysate.

Another model for the synthesis of prebiotic peptides was proposed only four years after the Miller experiment and was based on a reaction first reported by C. Krewson and J. Couch[28] in 1943. In August 1957, two months before the launch of Sputnik, the First International Symposium on the Origin of Life was convened in Moscow under the aegis of Oparin and his fellow scientists. At this meeting, Shiro Akabori[29] from the Institute for Protein Research at Osaka University presented a paper on a way in which polypeptides with molecular weights as large as 15,000 could have formed under primordial conditions. He circumvented the free energy dehydration problem by synthesizing polyglycine on clay from formaldehyde, ammonia, and hydrogen cyanide through aminoacetonitrile as the intermediate.

$$CH_2O + NH_3 + HCN \longrightarrow H_2N-CH_2-CN$$

Aminoacetonitrile

$$H_2N-CH_2-CN \longrightarrow (-NH-CH_2-\underset{\underset{NH_2}{|}}{C}-)_x \xrightarrow{x\ H_2O} (-NH-CH_2-\overset{\overset{O}{\|}}{C}-)_x + x\ NH_3$$

After polyglycine was formed, side-chains could be introduced by reaction with aldehydes or with unsaturated hydrocarbons. In this way, when Akabori treated polyglycine on kaolinite with formaldehyde, he converted 2 to 3 percent of the glycine to serine; acetaldehyde gave 1.5 percent threonyl groups. By this procedure he considered that valine, leucine, and isoleucine would be formed from propylene, isobutene, and but-2-ene; and reaction with acrylonitrile could lead to glutamine, arginine, and lysine. Other reactions could lead to cysteine.

$$
\begin{array}{ccccccc}
\mid & & \mid & & \mid & & \mid \\
NH & CH_2O & NH & & NH & H_2S & NH \\
\mid & \longrightarrow & \mid & \longrightarrow & \mid & \longrightarrow & \mid \\
CH_2 & & CH\text{-}CH_2\text{-}OH & & C\text{=}CH_2 & & CH\text{-}CH_2\text{-}SH \\
\mid & & \mid & & \mid & & \mid \\
C\text{=}O & & C\text{=}O & & C\text{=}O & & C\text{=}O \\
\mid & & \mid & & \mid & & \mid \\
\end{array}
$$

It was an ingenious scheme. Whether it is plausible for a prebiotic protein is hard to judge. J. D. Bernal[30] in 1951 had suggested that life may have originated on the surface of clay, which could have accumulated large amounts of organic substances. One difficulty with the reaction is removing the product from clay after it is absorbed. Akabori extracted his with dilute sodium hydroxide.

Clay is a common natural substance with structural and catalytic properties that have made it a useful material as an adsorbent and clarifying agent. The weathering of soils and different rock types leads to the formation of clay minerals as small particles that are essentially crystalline hydrous aluminum silicates. The minerals consist of combinations of two basic structural units: one unit is made up of silica tetrahydra with four oxygens or hydroxyls equidistant from a central silicon atom arranged as a repeating hexagonal network to form a sheet; in the other structural unit, two sheets are formed from minerals in which oxygen and hydroxyles are octahedrally coordinated around aluminum, iron, or magnesium atoms. Clays are grouped into illites, chlorites, kaolinites, montmorillonites, and others by their amorphous or crystalline features, as well as their swelling attributes. The ability of clay minerals to hold and exchange certain cations and anions is one of their most valued chemical properties.

Since there is ample evidence to indicate that amino acids could have been formed readily under prebiological conditions, an abiotic condensation of the amino acids themselves still has a logical appeal. But there remains the high energy requirement to polymerize them under natural circumstances. In order to do this, the amino acids are usually activated to a mixed anhydride between the amino acid and another acid. In biosynthesis the other acid is the phosphoric acid group of ATP and the resulting anhydride is the amino acid adenylate.

Once the adenylate is formed, the amino acid is elevated to an energy level high enough for polymerization to occur spontaneously. But for the amino acid adenylate to react with another amino acid, the amino group of that monomer must be without an electrical charge. Amino acids in neutral solutions exist as

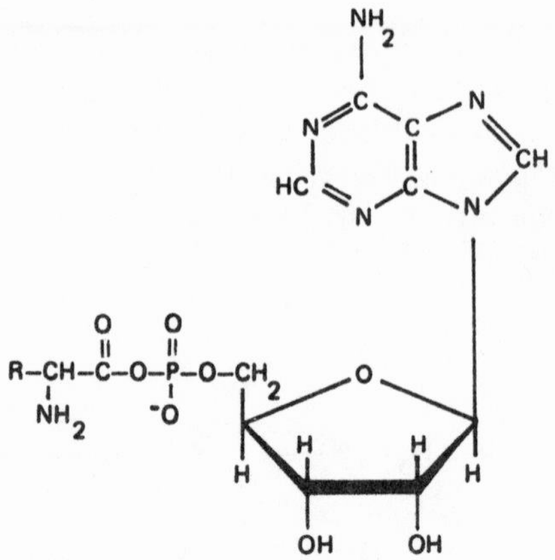

Figure 19.1. The structure of amino acid
adenylates.

zwitterions in which the amino and carboxyl groups are ionized. In order for either group to be reactive it must be without its charge, a condition that exists for the carboxyl below pH 5.4, for the amino group, above about pH 8.

If an amino acid and ATP are allowed to stand together at pH 7, no formation of the adenylate will occur. It will form in an acid solution where the un-ionized carboxyl group exists, but the amino adenylate hydrolyzes quickly at this side of neutrality. And too, the amino group cannot react unless the pH is around 8. What was needed for polymerization under natural conditions was a situation in which the carboxyl and amino groups of the amino acids would be reactive at the same pH.

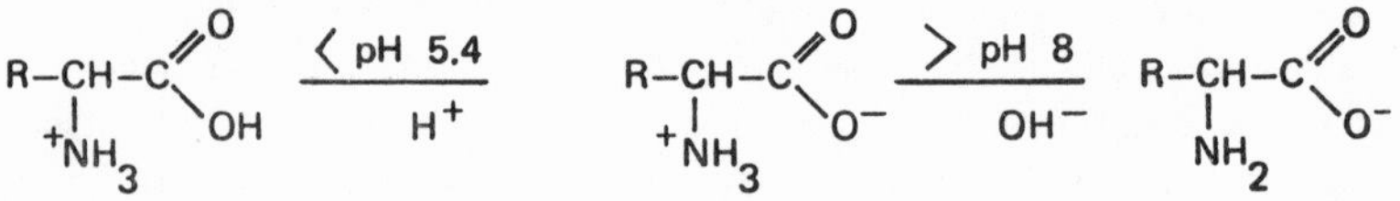

This condition was found by Mella Paecht-Horowitz and Aharon Katchalsky[31] at the Weizmann Institute in Rehoboth, Israel. They investigated clays, looking for the right silica clay that would have enough groups to combine with the amino acids to give a silicate salt of the amino group and free the carboxyl group from the zwitterion at neutral pH. Once free, the carboxyl could react with ATP to form the reactive mixed anhydride. They discovered that zeolites performed this function, and when used with a mixture of an amino acid and ATP, the adenylate was generated—but polymerization did not take place. Then, recalling their earlier success with montmorillonite, they attempted the experiment with this clay, and polymerization of the amino acid proceeded.

Under the experimental conditions, amino acids polymerized by being mixed with ATP and montmorillonite at pH 7.5 to 8.5. Because adenylic acid was liberated in the reaction, the pH had to be maintained by the continual addition of alkali; otherwise the polymerization would stop as the pH became acidic. The extent of polymerization achieved was impressive. Within a few hours, chains of 50 monomers and larger were attained.

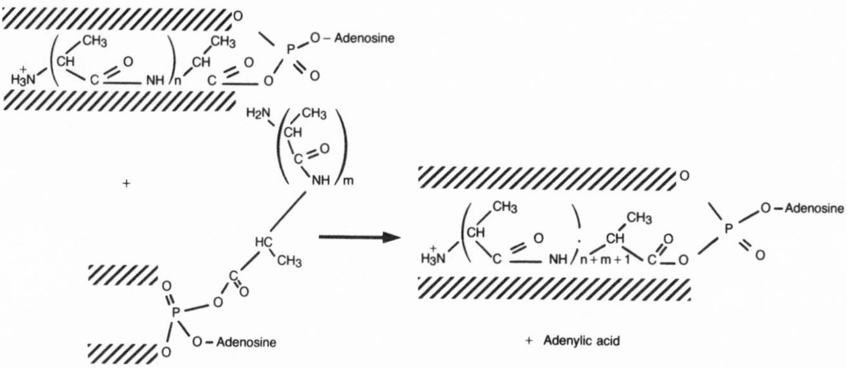

Figure 19.2. Amino acid adenylate polymerization on montmorillonite.

But there were drawbacks. Some amino acids like alanine and serine or alanine and proline copolymerized strongly, while a mixture of alanine and aspartic acid tended to give only homopolymers of each. And too, the basic amino acids do not polymerize because they are too strongly adsorbed to the clay.

An attractive feature of the reaction is that theoretically it could form the polyadenylic acid from the same starting material as a by-product of the reaction.[32] This has not been demonstrated experimentally, but if polyamino acids and polynucleic acids could be formed simultaneously as postulated by Eigen,[33] a principal hurdle in the solution to the origin of life would have been cleared.

In order for amino acids to condense on clay their concentrations have to be above 10^{-3} moles per liter. Most laboratory experiments have been carried out at relatively high concentrations in order to achieve optimal results, but the concentration of amino acids in the oceans may never have been very high. Klaus Dose,[34] by taking into account the production and destruction processes, calculated the concentration of amino acids in the primitive ocean to be about 10^{-7} molar. And N. Lahav and S. Chang[35] found the adsorption of amino acids at pH 8 too weak for these monomers to have accumulated on the surface of clays, as suggested by Bernal.[36] For this reason, they feel that any reaction on clay would have been preceded by a concentration step, such as evaporation in a lake or lagoon.

Clays have been found to be rare in pre-Cambrian sediments.[37] But though they may not have been abundant as the accumulation of clays from sedimentary rocks, clays nonetheless can result from the transformation of volcanic glass into argillaceous rocks. The hydrothermal action of volcanic fumaroles or weathering on volcanic ash and tuff produces mostly montmorillonite.[38] There would, therefore, probably have been clays present during the Archean to promote the polymerization of amino acids, as postulated by Paecht-Horowitz and her colleagues.

The clay-catalyzed reaction has merit but at the same time is objectionable. The greatest feature of the condensation reaction on clay is that it is an interfacial polymerization, a mechanism that is closely analogous to biological systems in which physical structure, such as surface-activated reactions between water–lipid interfaces, is so important. Furthermore, Armin Weiss[39] at the University of Munich has shown that inorganic clay systems can act as carriers of information by a process of spontaneous self-multiplication.

The most serious objection to the reaction is not the mechanism but whether it is applicable. The experiments by Paecht-Horowitz, Akabori, Fox, and others have attempted to demonstrate geological conditions under which large polypeptides could have been produced, apparently with the supposition that such a process could ultimately result in a prebiological synthesis of enzymes or enzymelike substances. And once polypeptides were present to catalyze the fundamental biochemical reactions, a functional living cell could assemble. It is doubtful whether anything resembling an enzyme could ever result from these reactions. But aside from that, these are deadend issues. The sequence of any utilizable polypeptide would not be retained and passed on to succeeding generations because it would not be from a replicating system.

20
The Enzyme Mystique

Enzymes have always had an aura of mystery. The extremely specific nature of their activity and their incredible rates of catalysis have made it appear that enzymes must be creations set apart from the commonality of chemical substances. But common or no, enzymes are nothing more than large molecules that have been honed by evolution to extreme precision and efficiency through 4 billion years of refinement.

We study contemporary life forms and are astounded by their biochemical complexity. Even the simplest bacterium has 2,000 to 3,000 different enzymes to perform its metabolic processes. Its enzyme-controlled cellular machinery functions so smoothly and efficiently that it has created the impression that enzymes in their present state were essential for the formation of any primal biological system. But we are looking at a far later system. Life did not begin with enzymes; it evolved with them through the eons of mutation and adaptation. To understand how a cell was able to start down the road leading to a living system controlled entirely by enzymes we need to know what makes an enzyme work. To understand enzymes we need to appreciate the character and singular significance of proteins, for enzymes are proteins.

Proteins, once they became a part of the biological systems, added a variability that permitted them to create the individual character of each living thing. There are tens of thousands, perhaps as many as 100,000 different kinds of proteins in every human body, representing more than half of the body's dry weight. They range in molecular size from 6,000 to as large as 10 million. Most, however, are between 12,000 and 36,000 (100 to 300 amino acids), the size of polypeptide chains usually synthesized by the ribosomes.

They serve in many capacities. As myosin and collagen, proteins are the structural component of muscle and connective tissue; as keratin, they form hair and nails; as enzymes, they control essentially every biochemical reaction; and as hormones, they are regulatory agents of cellular processes. They are antibodies that act in the immune response against alien proteins; and as histones and protamines, they suppress the expression of unneeded genes. Why is it that proteins have been given such a wide range of chores in the cells' operations?

The answer is that proteins can do what few other substances are capable of doing, namely recognize other molecules. The ability of proteins to identify molecules gives enzymes their highly precise specificity. The evolution of living systems to a large number of biochemical reactions depended upon the cell's ability to coordinate and exercise control over its many processes. This was made possible by the cell's evolving a particular enzyme to monitor each reaction. For this reason, not only are enzymes amazingly specific catalysts, but their number is extremely large.

How complicated is the chemical structure of an enzyme, and how could so many complex molecules of such precision have developed within the time of life on earth?

We know that all proteins heated in 6 Normal hydrochloric acid at 115°C for several hours are hydrolyzed to their subunits, the amino acids. The amino acids in protein are linked together in a long chain, with the order of their arrangement being primarily responsible for the protein's individuality. There are some twenty common amino acids (fig. 20.1), and they form the alphabet of protein structure. The structure is like a written language: the amino acids are the letters that form words; the words are arranged in sentences, which in turn become a fully expressed message. The message—the biological role of the protein—is spelled out in the total structure of the molecule.

The ability of proteins to distinguish between different molecules depends upon the proteins' three-dimensional shape. As life evolved and became more complex, the complexity of its proteins increased also. Biochemists realized that if they were ever to understand how proteins function in a cell, they needed to learn the precise spatial conformation of the molecules. This became one of the most formidable tasks ever encountered in the unraveling of the chemistry of life.

A polypeptide is a string of amino acids connected in sequence (fig. 20.2). When the molecule is large, however, as it is in the case of proteins, it is not linear but distorted by a number of factors into a spatial configuration. The variety of functional groups on the side-chains of the amino acids allows for interactions that stabilize some conformations and destabilize others. The sulfhydryl groups of two cystines can oxidize to form the strong disulfide bond; and the hydrogen donor groups such as $>$NH can form hydrogen bonds with the carbonyl groups ($>$C $=$ O) (fig. 20.3). The steric repulsion of bulky side-chains can prevent certain conformations; and water as a solvating structure around charged and polar groups exerts a strong influence.

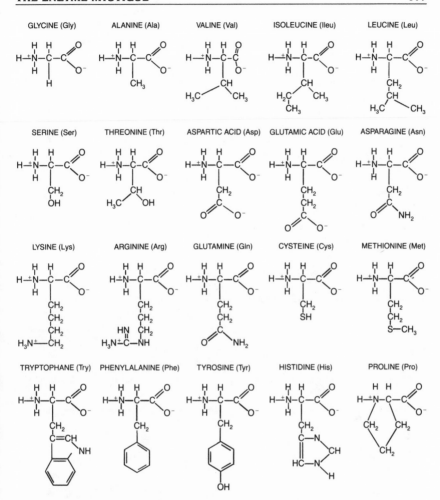

Figure 20.1. The twenty common amino acids found in protein.

In order to define the full three-dimensional shape and size of proteins, chemists have classed their structures in four stages. The primary structure of proteins is the amino acid sequence in the polypeptide chain. The other three stages are different spatial arrangements that the polypeptide chain assumes either alone or with other polypeptides.

Linus Pauling, Robert Corey, and their associates[1] at the California Institute of Technology discovered that the only way that asymmetrical building blocks of a polymer meshed so that each has the same relation to its neighbor is for them to form a helix, like a spiral staircase. The spiraling of the amino acid chain, therefore, becomes the secondary structure of protein conformation (fig. 20.4).

The third stage of the three-dimensional arrangement is the manner in which

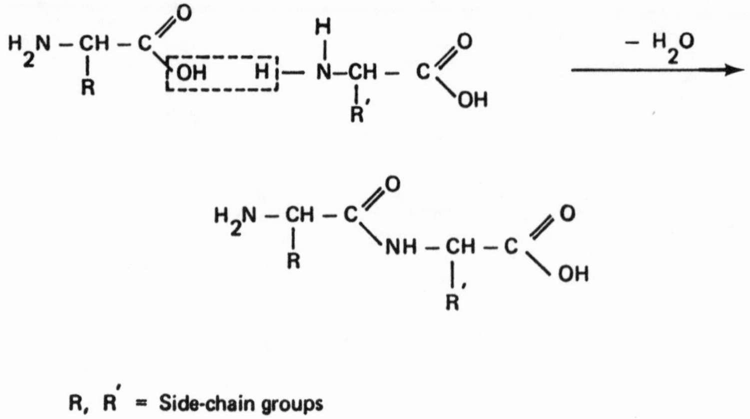

R, R' = Side-chain groups

Figure 20.2. A dipeptide.

the polypeptide chain turns, coils, and wraps around to a definite shape for the side-chain groups of particular amino acids to be precisely oriented. When the polypeptide chain is folded, it tends to push the polar groups to the outside, where they associate with water molecules, and to pull the nonpolar groups to the inside, where they form hydrophobic interactions. This leads to a spherical shape for the protein with a lipidlike interior and a polar and ionic surface (fig. 20.5).

There is a fourth level of organization to protein structure: the quaternary. Instead of a single polypeptide chain, several polypeptides, either identical or dissimilar, each with its own primary, secondary, and tertiary structures, can associate into an organized structure that has properties and specificities not exhibited by any of the monomeric units.

The awe-inspiring feature of enzymes that had been difficult to understand was how they could promote reactions so rapidly and for only one specific reaction. Emil Fischer, the German chemist who elucidated the nature of the peptide bond in 1902, pondered the question. When he discovered that yeast grows on the D-isomer of sugar but not on the L-isomer, he reasoned that there must be some

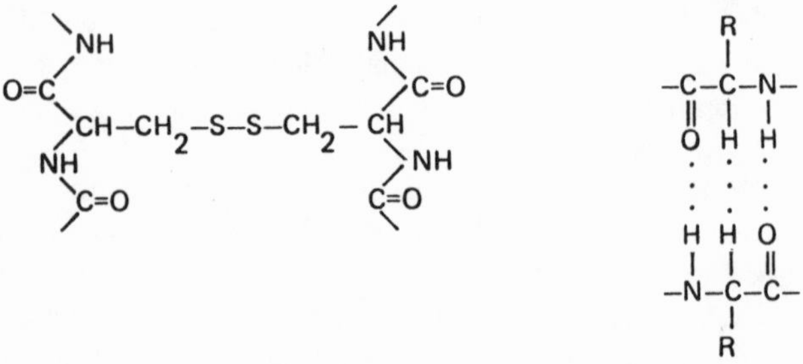

Figure 20.3. Peptides united by a disulfide bond, and hydrogen bonds.

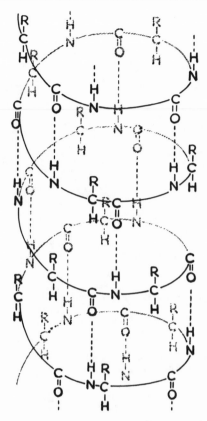

Figure 20.4. Alpha helix structure of a protein.

way in which only one isomer of the sugar molecule is able to fit into the enzyme, much like the way a key fits a lock. This became the lock and key hypothesis to explain the specific interaction between an enzyme and its substrate (molecules acted upon by enzymes).

Fischer's lock and key concept was elaborated upon by Leonor Michaelis and Maude Menton in 1913, when they hypothesized that the high specificity of enzymes results from the enzyme providing a site in its surface which a substrate molecule can bind to in a precise manner, forming an intermediary enzyme–substrate complex. Once the complex is formed, reactive groups in the enzyme promote the required chemical reaction on the substrate.

When protein chemistry advanced to the point where the amino acids in the active site could be identified in the primary sequence, researchers discovered that the participating amino acids were generally quite distant from each other in their position in the long polypeptide chain. It was now apparent why enzymes had evolved to such large sizes. It was out of necessity. Polypeptide chains are

Figure 20.5. Representation of primary, secondary, and tertiary structure of a protein (myoglobin). Dots represent the α-carbon of amino acids.

synthesized as a linear sequence of amino acid residues. But since the function of enzymes depends upon their three-dimensional shape for them to recognize a substrate and have groups oriented precisely for the active site, a chain of amino acids can accomplish this only by having a polymer long enough to be contorted into a globular shape.

Not until 1966 was the full three-dimensional structure of any enzyme determined. It was an enzyme first observed in 1922 by Alexander Fleming, the discoverer of penicillin. While working in London, Fleming came down with a cold. In his inquisitive manner, he used the occasion to add a drop of his nasal mucus to a culture of bacteria. To his astonishment, a few days later the bacteria nearest the drop of mucus were eaten away. Fleming recognized that the bacterial consumption was due to an enzyme, and since it could lyse, or dissolve, the bacterial cells, he named it lysozyme.

Lysozymes are widespread and have been found in several organs, in blood cells and plasma, in saliva, milk, tears, and egg white. The egg white lysozyme, whose primary structure was determined independently by Pierre Jollès and his colleagues[2] at the University of Paris and by Robert E. Canfield[3] of Columbia University College of Physicians and Surgeons, is a single polypeptide of 129 amino acid subunits. Since the lysozyme molecule may be readily unfolded and

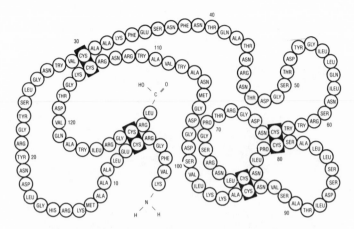

Figure 20.6. Amino acid sequence of egg white lysozyme.

refolded, not only its catalysis and specificity but also its three-dimensional structure is determined solely by these residues.

The function of lysozyme is to split a particular bond of a mucopolysaccharide, a long-chain complex sugar that is a component of bacterial cell walls. The cleavage is brought about by catalyzing the hydrolysis of a C-O bond. How lysozyme is able to cleave this specific bond became the research goal of an investigation by David Phillips and others[4] at the Royal Institute of London in 1960. The objective was to work out the three-dimensional structure of lysozyme in atomic detail using the X-ray crystallography method. In this method X-ray diffraction patterns are recorded and the image of the atoms in the molecule is obtained by calculation.[5]*

The lysozyme molecule contains some 1,950 atoms. The task of determining the three-dimensional structure of the enzyme demanded that all or nearly all of the positions of these atoms be established. In 1962, Phillips obtained a low-resolution image of the structure from 400 diffraction maxima and saw the general shape of the molecule. He had realized that his goal was not to be an easy one, but the arrangement of the polypeptide chain was even more complex than Phillips had anticipated.

Three years later, after the development of more efficient methods of measuring and calculating data and the use of nearly 10,000 diffraction patterns, an image was obtained. This time the resolution was high enough for many of the groups of atoms to be clearly recognizable. Phillips and his colleagues found that the giant molecule consisted of three sections of α-helix, two lengths of polypeptide chain running parallel to each other in opposite direction, a hairpin bend in the

---

* Microscopes for light waves or the electron beam can be focused, but no means of focusing X-rays into an image have ever been devised.

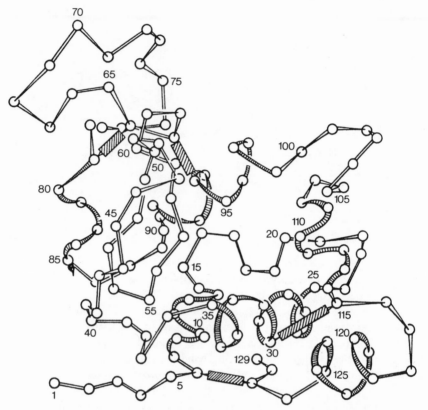

Figure 20.7. The lysozyme molecule. Disulfide bridges linking parts of the polypeptide chain are shown by diagonally hashed bars. The other hashed sections indicate the alpha helix.

chain, and other folding too irregular to describe briefly. The result of this coiling and turning of the polypeptide chain was a structure forming two wings lying at an angle to each other. The gap between the wings transformed into a deep cleft running up one side of the molecule. This cleft formed the active site of the enzyme and the cavity for the substrate to fit into.

By studying the interaction between lysozyme and eight different amino sugars, Phillips was able to show that the active site was bound by hydrogen bonding of six exposed amino sugar residues on the substrate. The many amino acids lining the cleft that are responsible for the bonding of the substrate and the enzyme are widely scattered along the chain. The more favored arrangement in which a model of the enzyme's substrate can be fitted into the cleft places the bond that is hydrolyzed directly between the carbonyl group of an aspartic acid in position 52 and the carboxyl group of a glutamic acid in position 35. Structural observations indicate that these two amino acids are the participants in the mechanism of action of lysozyme.

Lysozyme illustrates how a single polypeptide chain, by being sufficiently long and containing a variety of amino acids, can create a three-dimensional shape with a surface and orientation of amino acids that make it a catalyst of extreme specificity. The ancestral genes that translated to peptides in the development of enzymes mutated by doubling and substitution repeatedly, improving the old enzymes and introducing potentially new ones. Each type of enzyme of the 100,000 or so found in the human body did not develop independently, but all evolved from the few primeval genes that assisted in the formation of life. In this way the information for evolving entire sequences of amino acids was retained and used for making new enzymes as proteins became larger and more complex.

The mutation mechanism expanded the number and roles of enzymes, as well as of other proteins throughout the evolution of life. As the detailed chemistry of more proteins is determined by modern techniques, the process of evolution becomes more apparent. It has been discovered that the amino acid sequences of lysozyme and lactalbumin, one of the principal proteins of milk, are similar, but the functions are not. While lysozyme cleaves a β-1,4-glycosidic bond between amino sugars in bacterial cell walls, α-lactalbumin facilitates the synthesis of a β-1,4-glycosidic bond between glucose and galactose to form lactose, commonly called milk sugar. Lactalbumin has the general shape of the molecule with the active site conserved, but at the position of the catalytically active glutamic acid in lysozyme, lactalbumin has threonine. Apparently, the genes for lysozyme and lactalbumin resulted from gene duplication around 350 million years ago at the time of divergence of the amphibian line from that leading to the reptiles, birds, and mammals. The ancestral protein probably functioned like lysozyme today, but lactalbumin became more specialized and is found only in lactating mammary tissue.[6]

Enzymes have two molecular features that give them their astonishing powers: an active site that performs the catalysis; and a protein component that creates the physical shape for accepting specific substrates. A primitive cell emerging from prebiotic chemicals would have had neither the capacity to synthesize large polypeptides nor the precision of contemporary enzymes for specificity. The efficiency of enzymes would have developed by selection through evolution of proteins to greater size and refinement; the catalysis, however, would have been necessary to allow the cell to rise above the common mix of chemical interactions.

Enzymes consisting solely of polypeptide chains are a minority; most enzymes contain a coenzyme group in their active site that performs the catalysis. There are many substances—multivalent metal ions, imidazole derivatives, certain heterocyclic compounds—that catalyze chemical reactions. And large groups of enzymes contain these auxiliary components for their activity. These coenzymes are catalysts for particular types of chemical reactions but without the attendant protein component; they are not catalysts for particular organic molecules and lack the narrow specificity of enzymes.

Life began from an assemblage of substances in the primordial environment. There were no prebiotic proteins, but the emerging cells adopted in some manner a simple cellular machinery that started them on their road to synthesizing proteins to enhance catalytic proficiency. There is reason to believe that there were prebiotic peptides available,[7] and apparently biological cells began with catalysts complexed with relatively small peptides that were present in the environment.

Ferredoxin is the clearest example of an enzyme that began small and evolved in size and efficiency through the ages. This iron-sulfide protein is essential for the retention of energy trapped from sunlight and electron transfer. According to Dayhoff,[8] the ancestral ferredoxin may have been a complex of ferrous sulfide with a peptide of four amino acids consisting of alanine, glycine, aspartic acid, and serine. These are four of the most common amino acids formed in simulation experiments and are produced readily by heating ammonium cyanide.

21
Gene Splinting

A prebiological presence of large polypeptides may not have been necessary for the origin of a living cell, but the polynucleotides were definitely essential. A basic feature of biological systems that makes life possible is the replication of the nucleic acids and the translation of their structures to proteins. All attempts to devise reproductive systems not based on nucleic acids that could have preceded the biological systems remain unconvincing. It appears that the riddle to life's beginning will be solved only when a method is devised to demonstrate how the polymerization of nucleotides could have taken place under prebiological conditions on primordial earth.

There are four major nucleotide types: adenylic (AMP), guanylic (GMP), cytidylic (CMP), and uridylic (UMP) acids. Each nucleotide is a unit constructed of a purine and pyrimidine base, a 5-carbon sugar, and phosphoric acid. Without the phosphate group the nucleotide is called a nucleoside. In describing nucleotides, the hydroxyl groups of ribose are numbered to designate the position of attachment of the phosphate. In nucleic acids, the nucleotides are linked together in chainlike polymeric molecules by a phosphate bridge between the 3' position of the sugar of one nucleotide to the 5' position of the adjacent monomer. The nucleotides that form DNA differ from those that make up RNA by the absence of the hydroxyl in the 2' position of ribose, hence the sugar moiety is called 2-deoxyribose. Deoxyadenylic acid (d-AMP) and other DNA nucleotides are designated with prefix d.

The task for researchers in the origin of life studies has been to find some natural geologic condition in which the nucleotides could have joined end-to-end, creating prebiological nucleic acids. Leslie Orgel of the Salk Institute and Juan Oró's group at the University of Houston are among those who have been attempting to find an abiotic synthesis of polynucleotides under conditions simulating the environment of primordial earth.

As with amino acids, the polymerization of nucleotides is a dehydration reaction in which a molecule of water must be removed from each bond formed between two monomers. Unfortunately, the problem is rendered difficult by the relative unreactivity of the phosphoric acid to form the ester bond with the sugar hydroxyl. One cannot condense nucleotides merely by heating: pyrolytic destruction occurs before ester formation. Some nucleotides have been condensed using laboratory reagents,[1] but the problem remained to discover the natural circumstances on primordial earth under which abiotic polynucleotides could have formed. The solution to the riddle called for unraveling the way nucleotides could have become linked together to form genes of the first cells.

Nucleotides are not just ordinary chemicals. Formed by the joining of a purine or pyrimidine to a ribose and phosphoric acid group, these chemical units have singular properties that give them specific chemical behavior that the living cell has been built upon. One such property is that the mononucleotides can be activated to polyphosphate derivatives by further phosphorylation, in this way creating the high-energy phosphate carriers of chemical energy. But the uniqueness of nucleotides is that hydrogen bonding between complementary base pairs offers a built-in orientation mechanism for aligning nucleotides that has been the basis of their ability to replicate and to translate their structure through the genetic code.

It is this interaction of purines with pyrimidines through hydrogen bonding that is the molecular basis for much of the biological activity of nucleic acids. The hydrogen bond is of low energy—about one-fifth the energy of ordinary chemical bonds—but with chains of polynucleotides coupled in double strands the binding force can be large just from the accumulated bonding. And too, the base pairing is highly specific—guanine combines with cytosine and adenine with uracil. It is an intrinsic self-ordering associated with the chemical nature of the molecules.

The orienting of nucleotides in position before being condensed is an important step in both biological and nonbiological polymerization. When Kornberg and others[2] demonstrated that nucleic acids could be synthesized outside the living cell by using the enzyme polymerase and magnesium ions acting on the nucleoside triphosphates, they reported that the polymerization was very slow unless a primer nucleic acid was added to act as a template.

It was this self-aligning of nucleotides that Leslie Orgel and his group[3] at the Salk Institute tried to use to assist them to polymerize nucleotides in aqueous solutions in their search for the prebiological formation of polynucleotides. For nucleotides, they chose the monophosphates since they are more stable than the di- and triphosphates and would have been more prevalent on prebiological earth. To condense AMP, they added polyuridylic acid (poly U) to act as a template for the AMP to orient on. But since the phosphate group is not very reactive toward the sugar hydroxyl to form the diester linkage, they tried to handle this problem by using a water-soluble carbodiimide as a condensing agent.

Carbodiimides are laboratory reagents that complex with anionic groups (for

$$\text{example, } -\,\text{C}\overset{\displaystyle O}{\underset{\displaystyle O^-}{\diagdown}},\quad -\,\text{O}-\overset{\displaystyle O}{\overset{\displaystyle \parallel}{\underset{\displaystyle \underset{OH}{|}}{\text{P}}}}-\text{O}^-\,)\text{ to form activated derivatives which react}$$

readily with hydroxyls or amines to give the respective esters and amines.

The results of the experiment with AMP on poly U with the carbodiimide indicated that coupling of the AMP molecules occurred and was template directed, but the reaction was extremely inefficient. Furthermore, the dimers that formed were not of the 3', 5'-linkage found in nucleic acids, but a mixture of 2', 5'-and 5', 5'-bonded dinucleotides.

Orgel and his group tried another approach. Instead of using the relatively unreactive monophosphates, they decided to use activated derivatives of the nucleotides as the starting material. After all, they reasoned, when biological

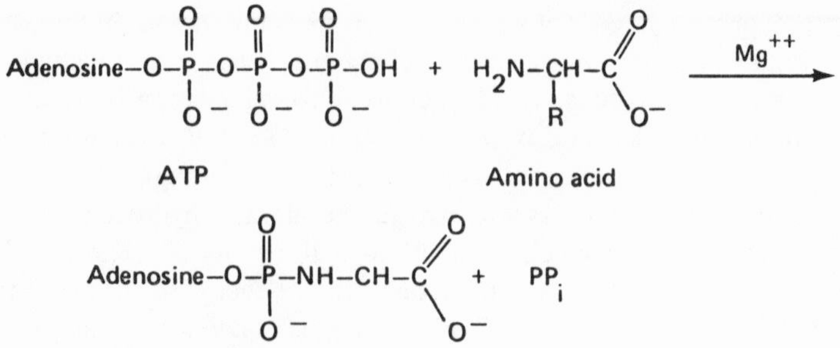

$$\text{Adenosine-O-}\overset{\overset{O}{\|}}{\underset{\underset{O^-}{|}}{P}}\text{-O-}\overset{\overset{O}{\|}}{\underset{\underset{O^-}{|}}{P}}\text{-O-}\overset{\overset{O}{\|}}{\underset{\underset{O^-}{|}}{P}}\text{-OH} \quad + \quad H_2N-\underset{\underset{R}{|}}{CH}-C\overset{\diagup O}{\diagdown O^-} \quad \xrightarrow{Mg^{++}}$$

ATP   Amino acid

$$\text{Adenosine-O-}\overset{\overset{O}{\|}}{\underset{\underset{O^-}{|}}{P}}\text{-NH-}CH-C\overset{\diagup O}{\diagdown O^-} \quad + \quad PP_i$$

Figure 21.1. A phosphoramidate.

systems make nucleic acids they use the triphosphate derivatives as the reactants. But organisms have an enzyme to promote the polymerization reaction. When ATP was mixed with the template in water containing magnesium ions, the ATP formed a stable helical complex with poly U but then hydrolized to ADP and AMP before any appreciable amounts of oligonucleotides (up to seven units) were formed. The nucleoside polyphosphates seemed too unstable in water to have been the answer, so the researchers began to look for some other type of activated nucleotide that could have existed on prebiotic earth that was more stable under the conditions than the triphosphates.

In examining the behavior of amines, including amino acids, when warmed with ATP and Mg^{++}, both in solution and in the dry state, one of the group, R. Lohrmann, discovered a novel series of reactions. Amino acids react with ATP when sufficient Mg^{++} is present to form phosphoramidates. Other amines, including ammonia, imidazole, and ethylenediamine, also gave phosphoramidates with ATP and other polyphosphates. The importance of phosphoramidates is that they are exceedingly resistant to hydrolysis in alkaline solution while still being activated at the phosphate group. Lohrmann[4] discovered that the imidazole derivative of adenylic acid, adenosine 5'-phosphoramidazole, condensed very efficiently on a poly U template to give dimers and higher oligonucleotides.

Thus the Mg^{++} catalyzed reactions of amino acids, imidazole, or other amines could have provided nucleotide derivatives that would have taken part directly in prebiotic polynucleotide syntheses. The activated intermediates, which formed best under dry conditions, would have formed in the solid state, and subsequently reacted in the presence of a small amount of water to form condensed nucleotides.

Renz, Lohrmann, and Orgel[5] explored the reaction further by using adenosine 2',3'-phosphate, the cyclic internal ester of adenylic acid. The cyclic 2',3'-

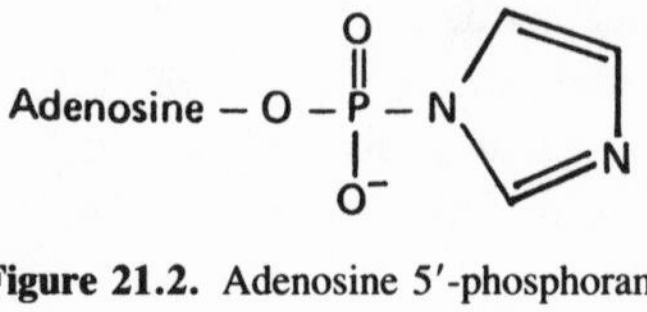

$$\text{Adenosine}-O-\overset{\overset{O}{\|}}{\underset{\underset{O^-}{|}}{P}}-N\diagdown\diagup N$$

Figure 21.2. Adenosine 5'-phosphoramidazole.

phosphates are the major product of the urea-catalyzed phosphorylation of nucleotides in the dry state and would have been a plausible chemical form of the nucleotides on primordial earth. Adenosine 2′,3′-phosphate did not condense on a poly U template with magnesium ions, but when the ions were replaced with simple catalysts such as glycinamide, ethylenediamine, or polyethylenediamine, good yields of the di- and trinucleotides were obtained. Presumably, the active intermediate in the reaction was the phosphoramidate formed by the nucleotide and the amine.

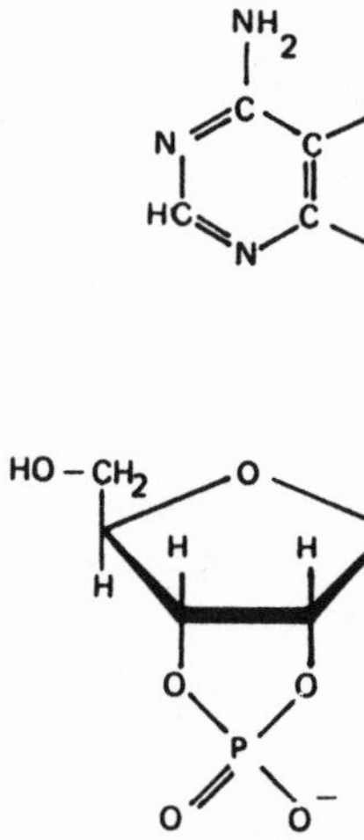

Figure 21.3. Adenosine 2′,3′-phosphate.

Until this time, the Orgel group was trying to find ways in which the nucleotides could have been polymerized in solution under conditions simulating the primordial oceans. For this the template was essential to bring together the monomers for reaction. This approach, however, would require an explanation of the prebiotic appearance of the template. Then Verlander, Lohrmann, and Orgel[6] discovered that under certain conditions the template was unnecessary. When the reaction mixture of adenosine 2′,3′-phosphate with the aliphatic amines at an alkaline pH was allowed to evaporate to dryness and incubate at 85–90°C, a self-polymerization of the cyclic nucleotide occurred. Oligomers up to the hexamer were obtained in good yields with 5 percent of the product larger than six units long. The length of the primal gene had become extended to six subunits.

The reaction using cyclic nucleotides was not limited to the adenosine derivative. When cytidine 2′,3′-phosphate was heated at 138°C for two days, Claude Tapiero and Joseph Nagyvary[7] obtained oligomers up to the hexamer in 50 percent yields. This is a high temperature, but just below 140°C, where slow decomposition begins. J. Skoda and his coworkers[8] had shown that uridine 2′,3′-phosphate underwent a similar polymerization.

While this was going on, Juan Oró and his group in Houston were attempting to polymerize nucleotides another way. As mentioned above, imidazole compounds appeared to be convenient reagents for the condensation of nucleotides. In 1969, O. Pongs and P. O. P. Ts'o,[9] two chemists at Johns Hopkins University, reported the polymerization of thymidine-5'-phosphate (d-TMP) to exclusively the 3',5'-linked oligonucleotides in a 40 to 50 percent yield using β-imidazolyl-4(5)-propanic acid as a catalyst. The reaction conditions, however, called for refluxing 15 minutes in the laboratory solvent dimethylformamide. Although the conditions were unrelated to a geologic environment, the catalysis and the results were similar to the desired effects of a simulation experiment.

The imidazole ring, as the side-chain of histidine, is an important catalytic unit in many enzyme reactions. Since acidic hydrolysis is initiated by a protonation step, enzymes that catalyze substrates by this mechanism have a proton donor and proton acceptor in their active site. The imidazole structure containing two nitrogens sharing a single hydrogen in a five-member ring with mobile double bonds is capable of acting as both proton donor and acceptor. The two nitrogens are actually equivalent, and with a shift of the double bonds, the hydrogens can reside on one nitrogen or the other. The imidazole has catalytic powers because it can donate the hydrogen from one nitrogen and recapture another hydrogen by the second nitrogen, thus regenerating the original structure in the process.

It is postulated that when imidazole catalyzes the coupling of the two d-TMP molecules, the exposed electron pair on its one nitrogen captures the hydroxyl hydrogen from the deoxyribose component, starting a chain reaction of shifting electrons.[10] The phosphate of d-TMP loses its acidic proton (H^+) on one side and its anionic oxygen breaks from it as it takes the imidazole hydrogen on the other to form the hydroxyl (OH^-). The simultaneous rejection of a proton and hydroxyl ion represents a loss of a water molecule and the two d-TMP monomers are joined in the process.

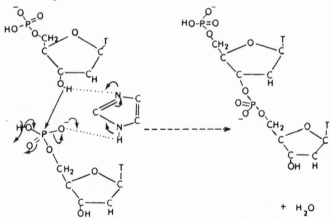

Figure 21.4. Postulated mechanism of d-TMP polymerization by imidazole.

In order to adjust the Pongs–Ts'o reaction as a model for the abiogen condensation of mononucleotides, J. Ibanez, A. P. Kimball, and J. Oró[11] attempted the reaction at elevated temperatures with d-TMP and imidazole in water. When the reactants were sealed in a glass tube and incubated at 90°C for 24 hours, the result was oligomers two to seven units long with the natural 3',5'-phosphodiester linkage. But the yields were only a few percent at the most.

The problem with attempting the polymerization of nucleotides in water is that removal of a molecule of water from the reactants is unfavored by mass action from the preponderance of the solvent. Nevertheless, there are chemical condensing agents that can extract water from compounds in an aqueous medium, and the Houston group investigated these. Carbodiimides, the type of compound that Orgel's group used at first, have been tools of organic chemistry for at least a decade, being used as convenient condensing agents. They were originally used in nonaqueous solvents because of their susceptibility to hydrolysis. But then it was discovered that the salt derivatives of dialkylamino carbodiimides,* which are water-soluble, react faster with anions than with water molecules. The mode of action for carbodiimides in condensation reactions is to react first with the anion to form a complex, which then hydrolyzes by extracting water from the monomers.† The large amount of free energy liberated by the hydration of the carbodiimides drives the condensation.

Carbodiimides are laboratory chemicals and would not likely have existed on primordial earth. The tautomer, on the other hand, is cyanamide,‡ a chemical that is believed to have been common when the building blocks of life were

* 1-Ethyl-3-3-(3-dimethylaminopropyl)-carbodiimide hydrochloride.

†

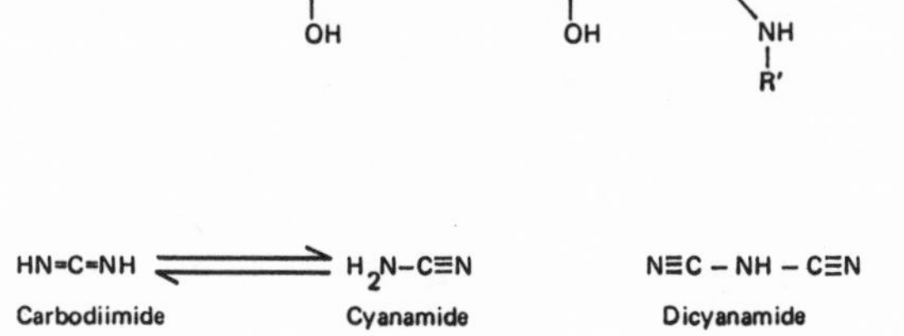

‡

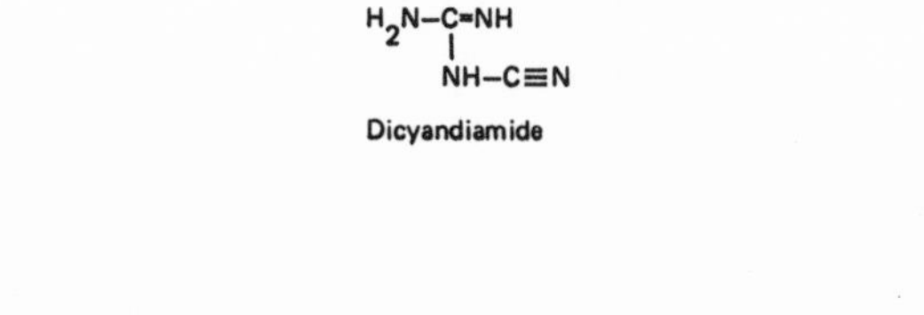

being formed. Cyanamide, its dimer, dicyandiamide, which forms in basic solutions, and dicyanamide, another cyanide derivative, are simpler condensing
agents than carbodiimides and could have been present and instrumental in
prebiotic condensation reactions. Cyanamide is formed by ultraviolet radiation
on aqueous solutions of ammonium cyanide. The reaction is not very efficient,
but the presence of halide ions improves the yield by a mechanism that is not
well understood. Considering the conditions of the reaction, it is conceivable
that large quantities of cyanamide could have been formed and were available
on primordial earth to act as a condensing agent.

Because the active species of cyanamide and dicyandiamide is a cation,* their
use as condensing agents is most effective in acid solution at pH 2. Such a low
pH may not have geologic relevance, but the reaction does occur with cyanamide
at higher pH's, although much more slowly. When cyanamide was used as a
condensing agent along with imidazole as a catalyst, d-TMP could be condensed
to oligomers up to five units in length.[12] But the results from solution reactions,
even with elevated temperatures, were not promising. The yields were low and
the prospects of extending the condensation reaction beyond a few units seemed
dim.

At this time it was discovered that one condition seemed to favor polymerization. Whenever the reaction mixture evaporated to dryness, the results were
invariably better. This gave a clue as to how the polymerization of the nucleotides
could have occurred—not in the primordial ocean itself but under the drying
conditions of evaporating lakes, the drying up of sea inlets, or even in the drying
and baking of solutions of the reactants splashed on rocks or mud flats on the
fringes of primeval basins.

E. Stephen-Sherwood and D. G. Odom, working with Oró,[13] extended the
condensation reaction of TMP by adding some of the triphosphate to the reaction.
They used cyanamide as the condensing agent, and for the catalyst they added
AICA, the imidazole derivative that was the intermediate in the synthesis of
purines from ammonium cyanide. When the reactants were subjected to drying
conditions and incubated at 60–90°C, oligomers up to four units in length were
obtained. The advantage of the mixture of nucleotides was that the yield of
oligomers was no longer small but actually 25 to 30 percent, or over ten times
greater than with the monophosphate alone.

Yet, despite the success of forming internucleotide bonds under conditions
similar to those of primordial earth, the extent of polymerization remained small—
too small to imagine assembling to form a functional cell. It was as though
Nature was jealously guarding the secret of her great event and had permitted

*

$$H_2N - C \equiv NH \qquad \overset{NH_2}{\underset{+}{H_2N - C - NH - C \equiv N}}$$

it to happen only after much difficulty. How then could polynucleotides have become large enough to cross the size threshold to perform in a functional cell?

The research on polynucleotides has been directed to finding conditions which would have promoted the polymerization of nucleotides into long chains to serve as primal genes. The results to date have indicated that oligonucleotides six or eight units in length could have formed under primordial conditions, but no one has succeeded in producing chains much longer. Nevertheless, nature succeeded in some way in creating polynucleotides long enough to eventually bring into operation the protein synthesis mechanism.

There is ample reason to believe that the chemical procedures used by biological systems were not chance developments, but evolved from the adoption of ordinary chemical interactions. The significance of this concept is that since life processes developed from one step to another in a continuum, it may be possible to retrace the path of evolution to detect what chemical reactions were the precursors to biochemical procedures used by living systems.

It appears that nucleotides were not polymerized readily into long chains. There must have been a system that permitted long chains to develop in some orderly manner. As pointed out earlier, the condensation seems to be assisted when it is template oriented, although it was found not to be a requirement under drying conditions. But it is quite possible that the initial long-chain polynucleotides came about on prebiotic earth in the same manner in which they are still being synthesized by organisms today—as the double-stranded molecule.

When contemporary cells copy DNA, they split the double-stranded molecule and condense nucleotides oriented on the single strands of the old DNA. It is unlikely that the cell evolved this method of duplicating DNA from its double-stranded molecule without precedence; rather, the duplication must have been a continuation of prebiotic chemistry. In other words, the prebiotic polynucleotides probably did not develop as single condensation products but as double-stranded molecules.

The problem leading to the original genes of the first functional cell was similar to that faced by H. G. Khorana and his colleagues[14] at the University of Wisconsin and the Massachusetts Institute of Technology when they sought a means of synthesizing a gene. In order to make the gene that codes for tyrosine transfer RNA, the research group did not attempt to polymerize a single linear molecule by condensing the monomers one after another. Rather they prepared 39 short fragments, each 10 to 15 nucleotides long. The segments were so chosen that a short piece of one oligomer extended from the other when two complementary fragments were allowed to form a double-stranded complex. The extended section then served as a splint to which an adjoining fragment could be complexed for coupling by the enzyme DNA ligase. In this way, adding first one strand, then another, the researchers were able to construct a complete double-stranded nucleic acid with an overall length of 207 nucleotides.

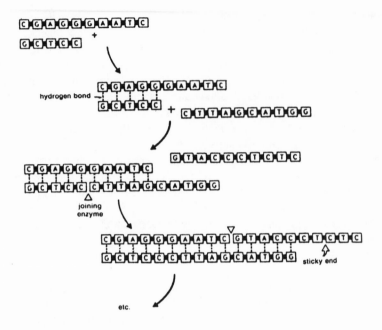

Figure 21.5. Scheme for synthesizing a gene using splint polynucleotides by the Khorana method.

This is how the prebiotic polynucleotides may have come into being. The results of the experiments simulating drying conditions on primordial earth, particularly in the presence of cyanamide derivatives as condensing agents, indicate that short oligomers formed readily from the nucleotides, but not long ones. With a variety of oligomers produced when primeval lake beds were flooded and evaporated repeatedly, it seems plausible that oligomers which would form the strongest complexes (that is, those with complementary base pairs) would be selectively retained. Whenever the adjoining segments fitted, they were in position to be coupled by the next drying cycle.

This scenario for the conditions that led to the first biological cells is consistent with our information from simulation experiments and could logically have occurred. To my knowledge an elongation of oligomers under these conditions has not been attempted in the laboratory. But it seems a reasonable explanation of how prebiotic polynucleotides formed on primordial earth—they grew, segment by segment, in a double-stranded molecule by a reaction repeated over and over until they crossed the threshold of functionality for a replicating cell.

22
"Particles of Life"

In the preceding chapters we saw the manner in which pre-biological peptides and polynucleotides could conceivably have formed on primordial earth. An accumulation of these chemical components set the stage for the final step—the assembling of a cell. Nucleic acids and proteins alone do not possess the character of life. They are static, whereas life is dynamic. A living cell is a working system, and it is the special organization of the macromolecules into a supramolecular structure functioning as an autocatalytic system that makes life. The biological unit then becomes a sequence of events where matter and energy are taken in to be converted to progressively more complex constituents of a self-sustaining structure.

To this stage experiments demonstrating the abiotic synthesis of biochemicals have been within the special interests of chemists, but the formation of a cell brings the origin of life to the level where it can be observed by everyone as an entity with the properties of microscopic organisms. In some manner the macromolecules that had condensed from the building blocks managed to associate and pass over the threshold to become life. They assembled into a coordinated arrangement that looked like and functioned as a cell. This was a quantum jump in the events leading to the formation of life and has, of course, because of its spectacular nature, received particular attention.

Essentially all substances have structural features that create associative forces with other matter, and they aggregate into an unlimited variety of amorphous combinations. But accumulations of organic matter do not just assemble into functional cells; they generally become undistinctive deposits. To form a cell, there had to have been a selection of molecules to fit together in a precise manner for the whole cell to be operational. The problem has been to find the procedure that nature took to assemble the prebiological components into an organized structure that became the living cell.

The most apparent necessity in the formation of the primal cell was containment. A living system cannot exist in free solution but only with the components confined in close juxtaposition. An envelope was needed to hold the constituents together while affording access to the surrounding environment for admitting raw materials and to allow a means of dividing during reproduction. Presumably, it had to be a surface that would isolate extremely small volumes of fluid from the surrounding volume of water. To pass from the inanimate to the animate, what substances did the primal cell find available in the earth's early geology to use for its cellular envelope?

A generation after Schwann discovered that the cell was the fundamental unit of life, men in the laboratory were attempting to create a living cell from common chemicals. But, like the alchemists and their dream of making gold from base metals, these researchers had no understanding of the underlying chemistry. As a result, models were designed primarily to duplicate the outward physical behavior of cells rather than chemical functionality.

In 1867, the German chemist Moritz Traube[1] carried out experiments in which he formed semipermeable films of copper ferrocyanide around small crystals of copper sulfate which he dropped in solutions of potassium ferrocyanide. These minute encapsulating spheres exhibited some features of growth, and Traube believed that they could be used to study the physicochemical properties of cells.

Another generation passed. Then in 1892, Otto Bütschli[2] prepared a model of a live cell by rubbing a drop of olive oil with potash solution. The chemically agitated droplet moved about in an impressive manner, engulfing particles in a fashion resembling amoeba. This experiment was carried further by others,[3] who created similar models that reenacted the movements, feeding, and division of cells.

The culmination of these pseudocells around the turn of the century was reached by the experiments of Stéphane Le Duc,[4] professor at l'École de Médicine de Nantes. Le Duc performed an experiment similar to Traube's by adding a piece of fused calcium chloride to a saturated solution of potash and potassium phosphate. A spherical shell of calcium phosphate formed, and as the calcium chloride continued to react, the osmotic pressure created within the sac caused it to expand. By changing the concentrations and adding various substances, Le Duc was able to generate exquisite osmotic forms that bore striking resemblances to algae, mushrooms, and other life forms. So enamored was the professor with his experiments that he believed he had embarked upon a new field of science, and he called it synthetic biology.

The followers of Le Duc's synthetic biology carried the experiments to conclusions even beyond the extravagent claims of the founder. It was an age when radioactivity was not well understood, and Madame Curie's radium was hailed as the wonder of science to rid the world of the ravages of disease. By using the mystifying energy of radium, Martin Kuckuck[5] converted a mixture of gelatin, glycerol, and salt into a culture of "cells" that were purported to have manifested all the signs of life.

Another generation passed before a realistic model for a primitive cell was devised. During the first half of this century, the chemistry of cells was believed to resemble the properties of colloids. A colloid is a special state of matter where particles are finely dispersed and suspended in a medium of a different phase. It can be solid or liquid particles suspended in a solid, liquid, or gas phase. The particles are small enough for their large surface areas to absorb ions and other materials to impart unique properties to the system. Electric charges within the substances or from absorbed ions cause the colloidal particles to repel each other and maintain the suspension. Familiar colloidal systems are smoke, fog, and airborne dust, as well as muddy water.

Whereas these are colloids that are generally of inorganic substances which lack an affinity for water and in which the particles remain dispersed by repulsion of their electrical charges, there are colloids of organic compounds that are quite different. These are hydrophilic colloids like gelatin, albumin, or starch that appear to dissolve in water but do not diffuse through a membrane.

Large molecules like proteins are water-soluble because water of hydration accumulates in structural masses around the charged and polar groups, holding the protein molecules in their uniform shapes by solvation. If the conditions are altered appreciably by heating, changing the pH or ionic strength, or by adding chemicals like urea that break up the water structure, the peptide chain becomes unfolded and adhesive. The protein molecules coagulate by clumping together and are no longer soluble (denaturation). Collagen, the fibrous protein of connective tissue and probably the most abundant animal protein, can be isolated as a crystalline compound, but once denatured, it becomes gelatin, the colloidal base of glue.

Since in the 1920s, proteins in the protoplasm were regarded to exist in a colloidal state, it was believed that the answer to the origin of life lay in demonstrating the manner in which a colloidal structure could arise from geological conditions. Information for such an experiment came from a phenomenon first observed in 1927[6] that was being investigated by Hendrick G. Bungenberg de Jong[7] of the University of Leiden, who reported on it in 1932.

Bungenberg de Jong had observed that hydrophilic colloids often spontaneously formed two phases. A fluid sediment in colloidal substances separated from and retained equilibrium with an overlying liquid layer free of colloids. He called the phase separation coacervation, and the colloidal particles coacervates. In some cases, the colloidal matter did not settle out from the liquid but remained as minute droplets suspended in the water. In effect, they were droplets of fluid separated from the surrounding medium by a thin film of hydrophilic colloids.

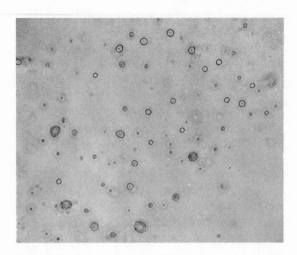

Figure 22.1. Coacervates prepared from gelatin and gum arabic.

The Russian biochemist A. I. Oparin, in his study of the origin of life, saw a similarity between coacervates and living cells. In his 1936 edition[8] he compared these colloidal droplets to the result when protoplasm is squeezed from plant cells. In spite of its fluid consistency, the protoplasm does not mix with water but remains as little droplets floating in the medium.

Oparin prepared various types of coacervates and studied their physical and chemical properties. Under controlled conditions they formed readily from proteins mixed with other select natural polymers. When a dilute solution of gum arabic, a commercial polysaccharide from the acacia tree, was mixed with a solution of gelatin at a pH under 4.8, coacervation occurred. Below pH 4.8, the isoelectric point* of gelatin, the charge on the protein became positive, while the gum arabic remained negative. The two polymers with opposing charges aggregated and formed droplets of the complex gelatin-gum arabic coacervate. By choosing pH levels between the isoelectric points of the polymers, Oparin was able to form a large number of coacervates with various chemical properties. Coacervates from gelatin and egg lecithin, from basic proteins such as histone and a nucleic acid, and other similar combinations were possible.

Oparin proposed that the phenomenon of coacervation could have served as the primitive protocell-forming mechanism. Coacervates do possess a boundary that segregates an environment from the surrounding medium, as well as other properties unique to the system; and they show many of the morphological characteristics of cellular inclusions such as vacuoles.

Coacervation can take place in extremely dilute solution—in concentrations as low as 0.001 percent;[8] and coacervates exhibit the property of colloids of

* The pH where anions and cations of a protein are equal.

being able to absorb substances. Coacervates prepared from gum arabic and protamine sulfate have been shown to absorb from a surrounding medium a bacterial lysate that is enriched in catalase activity.[9] And coacervates of gum arabic and histone or gelatin can concentrate starch from solution 3.5-fold. When β-amylase is present, it too is taken up by the coacervate and the enzyme hydrolyzes the starch. Numerous other experiments have been reported to illustrate that coacervates perform like primitive cells.[10] As an example, adding bacterial oxido-reductase to a coacervate mixture containing NAD_{red} created a simple anaerobic oxidation–reduction system.

Oparin reasoned that there were polymeric substances on prebiological earth that made coacervates that could have served as precursors to the first living cells. But essentially all the simulation experiments to demonstrate the cell-like properties of coacervates are with coacervates prepared from convenient biological polymers. Only if such experiments are performed with valid prebiological constituents are they meaningful simulation tests. And too, because of their colloidal nature, coacervates are easily susceptible to changes in pH and salt concentrations. Nevertheless, Oparin's coacervates were and have remained one of the most scientifically sophisticated hypotheses for the origin of the primal cell.

There have been other models proposed for the origin of the cell. In 1942, A. L. Herrara[11] of Mexico prepared what he called sulphobes by dissolving ammonia thiocyanate in formaldehyde and spreading the solution in a thin film. After several hours, microscopic structures with activities analogous to those of living organisms were reported to have been produced. Herrara saw in these minute shapes the varieties that are formed with amoeba and tissue forms and believed they may have been significant in simulating the origin of life. He pointed out that the conditions in volcanoes would have produced ammonium thiocyanate by a known reaction from sulfur.

It was not until after the Miller experiment in 1953 that other researchers offered models of cellular and precellular organization. Karl Grossenbacher and C. A. Knight[12] at Berkeley, while carrying out research with the sparking experiment, isolated not only amino acids, but also peptides. Then, in examining the solid film and samples of the cloudy liquid from the reaction mixture, they discovered small, solid bodies which range in size from about 0.08 micrometers down to 0.005 micrometers and less. These spherules, in the size range of viruses, contained large amounts of minerals, but also amino acids and similar compounds. The authors suggested that the spherules were silicates stemming from the borosilicate glass of the sparking apparatus and doubted if they played any role in biogenesis.

A research group in Romania investigated the possible formation of organic compounds from a primordial atmosphere under different circumstances. Simionescu, Dénes, and Macoveau[13] considered prebiotic components forming by electric discharge in cold plasma conditions adjusted between $+20°$ and

− 60°C. In simulation experiments under these conditions, a mixture of methane, ammonia, and water vapor gave evolved organic structures. The results were complex polypeptide substances which, when viewed under the microscope, were seen condensed in fibrillar, branched, or circular proteid arrangements.

The fact that the Miller experiment made formaldehyde and thiocyanate plausible prebiotic substances placed a new relevance on Herrara's sulphobe structures. Adolph Smith of Sir George Williams University in Montreal with J. J. Silver and Gary Steinman[14] at Pennsylvania State University reinvestigated Herrara's claim of observing lifelike forms in his reaction mixture. The researchers discovered that when the ammonium thiocyanate and formaldehyde were mixed, the colorless solution became slightly red within seconds, and finally a golden yellow after one hour. Microscopic examination revealed a high density of spheres 1–5 micrometers in diameter. Additional experiments showed that ultraviolet light promoted the reaction. And when zinc chloride was added to the reaction mixture, the zinc became incorporated in the structures, bestowing a localized ATPase-like activity. The chemist felt that the experiments clearly indicated how simple chemical substances that are believed to have existed on primordial earth could have been implicated in the formation of cell-like structures before any living cells existed.

Another model for precellular organization that has been much publicized is spheres formed from thermal polymers of amino acids by Sidney Fox and his collaborators. In 1959, Fox, Harada, and J. Kendrick[15] reported that hot, saturated solutions of their acidic proteinoids, upon cooling, gave large numbers of uniform microscopic spherules. These spherical structures, called proteinoid microspheres, were usually 1.5 to 3 micrometers in diameter and were prepared

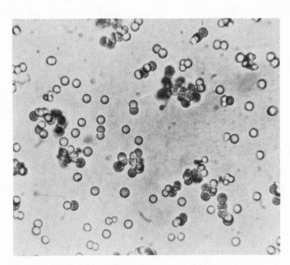

Figure 22.2. Microspheres formed from the acidic amino acid thermal polymer (proteinoid) boiled in water.

readily from the acidic proteinoid. Basic proteinoids, on the other hand, did not normally make microspheres except when mixed with the acidic proteinoid. Microspheres containing basic proteinoids gave a positive Gram stain; those prepared from acidic proteinoids took up the stain but were negative. This observation was taken to indicate similarities of surface composition in proteinoid microspheres and bacteria.[16]

As is the case with most of these physical models of cells, the spheres can coalesce and divide, and smaller members can adhere to larger ones. This latter feature, because of its similarity to live microorganisms, has been referred to as "budding." Removing these smaller spheres with mechanical, thermal, or electric shock and collecting them by centrifugation, Fox and his group reported that they observed them grow in size when placed in a saturated proteinoid solution at 37°C.[17] And changes in the solution conditions induced these microspheres to undergo various morphological changes simulating replicating cells. In all, five means of "reproduction" have been attributed to these nonbiological spheres.

Like Le Duc with his osmotic bags of seventy years earlier, Fox has tried to correlate all the bubble physics of proteinoid microspheres to attributes of living cells. By applying a slight pressure to microspheres, they can be made to associate in chains like algae;[18] when spheres fused and there was an exchange of occluded material, it was called communication;[19] and the most recent interpretation of conjugation between microspheres has been declared to be protosexuality.[20]

It is difficult to see that proteinoids or their microspheres have any relevance to the origin of life. As J. D. Bernal pointed out, "such spherules are quite commonly produced by irregular or branched-chain polymerization, as in starch grains, which cannot effectively lead to a crystal, but forms an assembly of rigid or plastic spheres around nuclei. Any resemblance to organisms, such as the presence of double spheres indicating fission, is probably fortuitous."[21] In effect, Fox's research reflects a retrogression to the older experimental approach to the origin of life, which was aimed at producing imitations of primitive life shapes.

Certainly microspheres represent an additional indication that the thermal polymerization of amino acids does not result in copolymerization resembling natural proteins. If one considers the acidic proteinoids as being primarily polyaspartic acid structures, the intermolecular bonding is not difficult to explain. Carbozylic acids have a strong tendency to dimerize by joining through their carboxyl groups. In the same manner, the aspartic acid side groups in the proteinoid form intermolecular bonds with each other in adjacent molecules. This is consistent with the fact that microspheres dissociate above pH 6 due to ionization of the carboxyl groups.

By changing the salt concentration, one can make microspheres swell and shrink, a property that has been suggested to be similar to that which cells exhibit when they are subjected to a change of osmotic pressure on their semipermeable membrane.[22] Since microspheres have been demonstrated to be too leaky to act as a membrane,[23] the swelling and shrinking cannot be due to osmotic pressure,

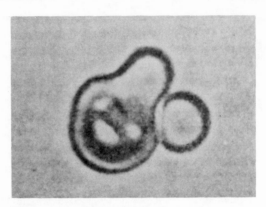

Figure 22.3. *Jeewanu* particle showing "budding."

but is obviously from the proteinoid acting like a charged polymer in the manner of Dowex 50.[24] Polymers containing ionizable functional groups (for example, $-C{\underset{\smallsetminus OH}{\overset{\displaystyle =O}{}}}$, $-NH-$) shrink or swell in varying salt concentrations due to the change in the number of charges on the polymer. High concentrations of salt create shrinking by electrostatic attraction of the charged polymer to the opposing ion. Such polymers are commonly used in the laboratory as ion exchangers.

Approximately two dozen properties of proteins have been assigned to the thermal polymers of amino acids.[25] The "limited heterogeneity" and nonrandomness of the composition of proteinoids have been regarded as supporting evidence that amino acids can be thermally copolymerized in sequences resembling proteins. This, however, is inconsistent with the physicochemical behavior of proteins. Proteins and polypeptides do not form microspheres. Even denatured proteins, as when an egg is boiled, do not aggregate into microspheres, but coagulate in the familiar manner.

Similar efforts to create models of the primal cell have given greater regard to the gross morphology than to chemical functionality. Krishna Bahadur[26] of the University of Allahabad, India, in 1964 prepared microspheres from para-formaldehyde, water, and colloidal molybdenum oxide irradiated with visible light. It was claimed that these microscopic spheres were capable of growth and division with metabolic activities. Bahadur also prepared microspheres by the thermal polymerization of amino acids and called them *Jeewanu,* the Sanskrit word for "particles of life."[27]

The rationale put forward by Oparin, Fox, and others for their models is that the living cell was preceded by a system that had morphological properties resembling cells but was not yet living. Purportedly, Oparin's coacervates became protobionts,[28] and Fox's microspheres, protocells.[29] These structures then supposedly underwent a period of evolution in which they changed until they evolved into the first living cells. No matter how you look at it, this is scientific nonsense. Evolution is a biological process of development through mutation, reproduction, and selection. These pseudocellular models, like clay, soap bubbles, or other inanimate objects, have neither the mechanism nor the potential of becoming anything more than what they are.

Such models for living cells lost their scientific value when the purpose of the cellular membrane was forsaken to emphasize impressive morphological similarities. The hypothesis of coacervates was founded in an age when colloidal chemistry appeared to be the basis of life. Proteinoids arose from an unsuccessful attempt to synthesize an abiotic protein or proteinlike material by a simple process. Both hypotheses were overextended to try to encompass the role of prime mover in the origin of life, despite the tremendous strides in molecular biology and membrane chemistry which revealed that the chemical principles upon which they were derived had little similarity to the basic chemistry of biological systems.

The question at stake is what forces would have been responsible for the association and assembly of the prebiological polymers. The principal binding force of coacervates is the electrostatic attraction between hydrated biopolymers; the formation of microspheres is dependent basically on the same interaction. Coacervates are generally prepared from biological compounds, and microspheres from nonbiological compounds. The occurrence of the former on prebiotic earth has had to be justified; and the latter are dead-end substances that would not be synthesized by a functional cell.

But the most serious fault of models based on particles held together by ionic forces is that they would have been continually periled with dissolution. Coacervates are notoriously unstable, and microspheres exist only in saturated solutions. Their existence in Archean lakes or oceans would have been short-lived.

23
The Vital Envelope

At some point on primordial earth something happened, something that was commonplace and spontaneous in the stages developing toward a biological system that marked the beginning of life, for it was from that moment on that certain prebiological peptides, polynucleotides, and assorted substances were no longer just chemicals but biochemicals associated in such a way that they became a living cell. What was the nature of that final step? It must have been more than the generation of bizarre spherical particles that are substantively unrelated to biological systems. Rather, it must have been a process that introduced a fundamental property to the organization of matter that distinguishes a living cell. That critical event that became the moment when life began was the encapsulating of a gathering of self-replicating molecules within a semipermeable cellular membrane.

A living cell became a special interaction of matter intermediate between substances firmly bound by covalent bonds and molecules in free solution. In order for the structure of polynucleotides to be expressed through replication and translation, they and their auxiliary constituents had to be confined in close proximity, but still with a degree of freedom to move and interact. For this the cell needed an enclosing sheath that possessed all the physical attributes to segregate the cell from its surrounding environment, while still allowing selective access to raw materials and energy. In this way it could maintain an energy potential with respect to the surroundings to drive its cellular machinery.

The confining envelope needed to be impermeable to the cellular components but able to allow admittance of small molecules; it needed to be elastic for expansion and division, but held together spontaneously; it needed to be insoluble in

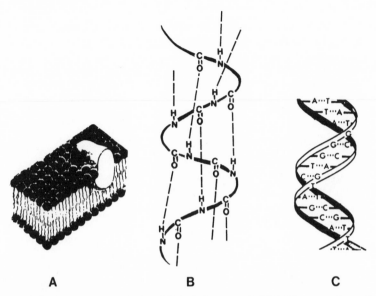

Figure 23.1. Basic structural units of biological systems. A. Lipid bilayer membrane. B. The helical coil of proteins. C. The double helix of nucleic acids.

dilute solutions and over a broad pH range; and it needed to be composed of simple chemicals that seal together automatically into the semifluid, two-dimensional conformation of the envelope. All these requirements were fulfilled by the bimolecular lipid membrane.

The lipid bilayer membrane is a model of simplicity that forms spontaneously and is a structural feature universal to all living things. Every cell, including the smallest free-living microorganism, has it. So fundamental is the lipid bilayer that it joins the α-helix of proteins and the double-helix of DNA in representing the three basic structures of biological systems. The importance of the lipid membrane is indicated by the fact that the more complex organisms have become, the more extensive has become the number of cellular structures containing membranes.

The composition of lipids in membranes of contemporary cells varies considerably, but generally a membrane consists of approximately half protein and half lipid, with the lipids being mostly phospholipids. These latter compounds represent a series of fatty acid derivatives of glycerophosphoric acid that vary by the small molecular component joined to the phosphate group.

Because the hydrocarbon chain of fatty acid substituents is repelled by water, whereas the phosphate end of the molecule is polar and water-soluble, phospholipids are a class of chemicals that align themselves between two physical phases. In water, the phospholipids are driven together by the exclusion of the paraffin chains from water into aggregates as sheets of molecules with the polar ends

Figure 23.2. Micelles and liposome (vesicle).

extended into the water. As a surface film, the aggregate is a monomolecular layer at the air–water interface with the hydrocarbon chains directed outward away from the water. Beneath the surface, however, surrounded by water, the phospholipids form spherical shapes with the hydrocarbon chains fused with each other. If a single layer of phospholipid molecules fuse their hydrocarbon chains to form a ball, a micelle results; but if a bimolecular layer creates a spherical enclosure, the particle becomes a liposome, or, when it refers to biological components, vesicle. It is the vesicle that represents the ubiquitous cellular feature of all life on earth.

A liposome formed of the closed bilayer sheet creates the condition of the living cell where the internal environment can be maintained at a different chemical composition and energy potential from the surrounding medium. The reproductive and metabolic machinery of a cell could not function unless the substances of low molecular weight that are substrates and precursors for biosynthesis were retained within a membrane. These double-lipid membranes with their hydrophilic ends outward are intrinsically stable barriers to prevent the loss of valuable cellular constituents and to maintain the concentration gradient essential for a biological system.

The membrane has to isolate the cell from its environment—but not completely. Nutrients must enter the cell and metabolic wastes must be expelled. Generally, a cellular membrane is pervious to lipid-soluble substances and water, but admits only soluble organic compounds containing no more than three to five carbon atoms. The penetration of ions depends on the number of electric charges per ion; polyvalent species such as calcium and sulfate ions appear to be very slow to penetrate cells. And there is a barrier to the positive charge; small anions like chloride and bicarbonate penetrate the membrane of red blood cells about a million times faster than cations of the same size. This selectivity of ionic charge, characteristic of the cellular membrane, is also demonstrated by liposomes of phospholipids by several orders of magnitude.[1]

The protein portion of contemporary cellular membranes contains enzymes and other specific molecules that perform functions for today's organisms that would have been too advanced for the unsophisticated encapsulating envelope that the primitive cells possessed. There are, however, some processes that, because of

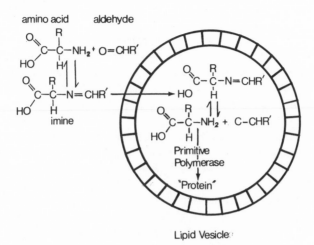

Figure 23.3. Scheme indicating the aldehyde mediated uptake of amino acids into a lipid vesicle.

their fundamental character, would have been shared by the primal cells and their more evolved descendants alike. Primarily, all would have had to bring into the cell across the lipid barrier nutrients from the outside medium to fuel their growth.

Many substances of importance—sugars, amino acids, nucleotides—are not soluble in lipid, exceed the size limit of permeability, and yet penetrate cellular membranes easily. It is believed that contemporary cells possess a special transport mechanism to take these essential materials across the lipid barrier. Since this active transport system appears to be constructed of special proteins in the membrane, it is too developed to have been available to the primal cells. How then would vital components have crossed the lipid layer to supply the primitive cell from the outside medium?

Some simple mechanism available to the lipid vesicle must have existed for the cell to accumulate small, water-soluble molecules. William Stillwell[2] proposed a mechanism that could have served as a rudimentary transport system for the diffusion of amino acids, sugars, and nucleotides across the lipid membrane of primal cells. When amino acids are condensed with simple water-soluble aldehydes, the resulting imines are capable of diffusing across the lipid barrier. Once inside the cell, dissociation of the imine would liberate the amino acid, which would be retained by the membrane; while the aldehyde, capable of diffusing across the membrane, would pass out again from the cell. Similarly, sugars could have crossed the membrane condensed with amines,[3] and nucleotides could have penetrated into the lipid vesicle as monovalent and divalent metal ion complexes.[4] This type of diffusion is known as carrier mediated, or facilitated, diffusion. Since it does not involve complex transport proteins, it could have been an important transport system with the primitive cells.

The lipid bilayer membrane possesses an additional property, both unique and indispensable, for the developing cell. It gave the cell its insulating capacity to electric charge. All membranes of living cells manifest a difference of electrical potential between the two sides of the membrane. In order to absorb photons of sunlight and pass the energy along the electron transfer system to synthesize ATP, the chain of electron carriers has to be insulated to prevent short circuiting. Only the lipid membrane possesses the thin, insulating features necessary to perform as a biochemical capacitor.

The lipid vesicle appears to be the sole model for the envelope of a primordial cell that fulfills all the biological requirements. Coacervates, proteinoid microspheres, Jeewanu particles, and inorganic spherules do not have all the vital properties of a lipid bilayer membrane. They are too leaky to retain small molecular weight substances, too thick and conducting to serve as insulators, and they all fail to comply with the principle of continuity, the generalization that requires each stage of evolution to develop as a continuation from the previous one. There have been efforts to incorporate essential features of the lipid membrane into coacervates and microspheres, but the purpose becomes suspect. Neither coacervates nor proteinoids have any role in biological systems, nor is there evidence that they have ever had. On the other hand, the lipid bilayer appears to have been the vital envelope from the beginning of life.

The creation of a biological system was the result of natural substances responding to spontaneous reactions. The building blocks were found to be energetically favored structures produced from the interaction of intense energy with matter. Their polymerization follows set chemical reactions under common terrestrial conditions; the nucleotides form pairings through hydrogen bonding for replication; and polypeptides wrap themselves into the helical coil. So too, then, do lipids assemble spontaneously in an aqueous environment to form the basis of the cellular membrane.

The spontaneity of lipids to structure formation is illustrated by a drop of oil on water. Even a minute amount of oil will automatically spread into sheets of aligned molecules, forming a thin film. Surface films are monomolecular layers formed at the interface of water and air, whereas, within the aqueous volume without the interface, the lipids in a sense create their own interface by keeping their polar ends toward water, while orienting their nonpolar tails together. If a sphere is formed encapsulating water, the result is a bilayer membrane enclosure—a vesicle.

This is an ordinary feature of nature whose significance to the origin of life was first pointed out by Reg Goldacre[5] of the Chester Beatty Research Institute of London. In 1958, Goldacre published a paper on surface films that he observed to be common to all lakes, rivers, and streams. Rarely noticed unless dust-ladened or examined carefully, a monolayer of elastic film is present on the surface of all bodies of water. It appears to be a part of the natural environment. Tadpoles eat it, freshwater snails crawl upside down along the under surface, and small arthropods can be seen supported by it as they hop across the surface.

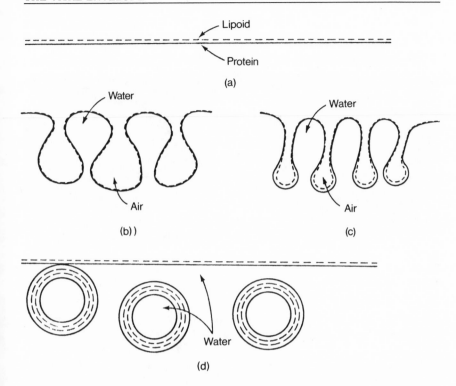

Figure 23.4. Proposed model for the formation of biphasic vesicles by wave action on surface films.

It is a molecular film, not of oil, but presumably of lipoproteins from leaves, pollen, and other material that fall in the water.

When the film is compressed by a reduction in the surface area, as occurs when a stream changes abruptly from shallow to deep, or by the flow of water under a floating barrier, or even by surface pressure from the wind, the film wrinkles in a series of folds. The buckling film, in cases where the hydrophobic surface folds back upon itself, fuses, entrapping air and water in alternating loops. As the air gradually escapes, the bubbles collapse, bringing together adjacent sections of the film to form a double lipoprotein membrane which severs from the surface film in little cylinders that roll away freely in the water. These elongated structures quickly give rise to spherical cell-like particles, each consisting of the lipid bilayer membrane with an enclosed aqueous environment. Goldacre found that the properties of collapsed lipoprotein films—surface tension, adhesiveness, osmotic expansion-contraction—resemble closely the properties of biological membranes.

Lipoproteins forming the surface films are complexes between proteins and lipids that are principally phospholipids. It is the hydrophobic paraffin chain of the fatty acid components of the phospholipids, being excluded from water, that

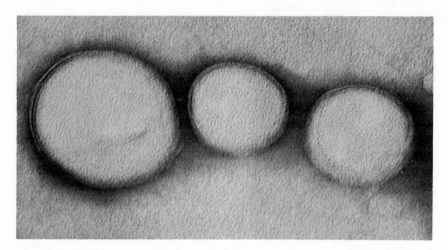

Figure 23.5. Vesicles produced in an extract of an experiment simulating the prebiotic synthesis of lipids.

orients the lipid molecules in a monomolecular layer at the air–water interface. The binding between protein and phospholipid is generally believed to be predominantly ionic and at the interface, although hydrophobic side groups on the protein can interact with the lipid layer. When a bilayer results from the collapse of the film, the hydrophobic chains of the two layers of lipids are fused and the pinched off unit becomes a vesicle with the proteins attached to both sides of the membrane.

The surface film is of biological origin. Would its counterpart have existed on prebiological earth?

It seems probable that it would have been a common feature then, as it is now. Phospholipids are compounds that result from the condensation of fatty acids, glycerol, phosphoric acid, and an additional small component. It has been reported that cyanamide, the condensing agent implicated in the abiotic formation of polynucleotides and polypeptides, also catalyzes glycerol phosphorylation in acid solutions. In experiments studying the abiotic synthesis of phospholipids on primordial earth, Will Hargreaves, Sean Mulvhill, and Dave Deamer[6] of the University of California at Davis found fatty acids and fatty aldehydes reacted with glycerol when mixtures were dried together and incubated at 65°C for one week. Then, in experiments simulating tidal pools, they prepared dilute solutions containing glycerol, phosphate, and cyanamide and added various hydrocarbon derivatives for the surface film. When they evaporated the mixtures to dryness and baked the residue at 65°C on sand or clay, they were able to detect in the resulting material the presence of phospholipids similar to phosphotidic acid and phosphatidyl glycerophosphates, and some residual neutral lipids. Moreover, when water was added to the products and the mixture was agitated, a hetero-

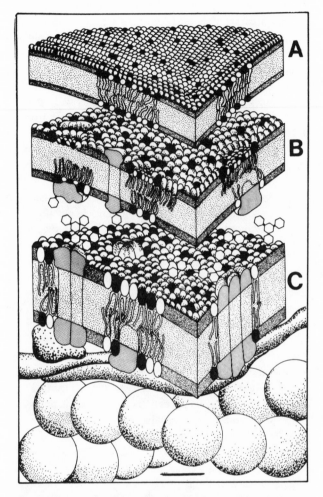

Figure 23.6. Hypothetical steps in biomembrane evolution. The bar represents 50 nm. A. Simplest bilayer membrane containing single-chain lipids with C_8-C_{12} hydrocarbons. B. Membrane containing both monoalkyl and dialkyl lipids (C_{10}-C_{16}) with adsorbed and intercalated polypeptides. C. Advanced membrane of C_{16}-C_{18} dialkyl lipids containing proteins and carbohydrates, and associated with peripheral protein systems.

geneous suspension of vesicles with the lipid bilayer membrane formed readily. A particular significance was attached to the phosphatidyl glycerol because this lipid and its aminoacyl derivatives are prominent in membranes of procaryotes.

As Hargreaves and Deamer[7] continued their study, they found that the tendency to form cell-like structures was exhibited not only by lipids simpler than the phospholipids, but also spontaneously without wave action or other forms of turbulence. Single-chain charged molecules formed stable lipid vesicles, and even the uncharged monoacylglycerols coalesced into liposomes when dispersed in water. Furthermore, saturated fatty acids of 8 to 16 carbons formed lipid vesicles when the conditions were slightly alkaline.

The critical parameters for the formation of vesicles were pH, temperature, and the hydrocarbon chain length. The minimum chain length from which membranes could form in pure dispersions appeared to be 8 carbons for both fatty acids and monoacylglycerides. The required temperature range was between 20° and 55°C, and the pH, 7 to 9, conditions that exist within most terrestrial environments.

The simple glycerolipids were presumably more prevalent than phospholipids on the primordial scene. Sand or clay in dry mudflats or catchbasins could have provided the circumstances for condensation-dehydration reactions where the hydrophilic and hydrophobic molecules combined to form glycerolipids. In addition to surface catalysts such as silica and clay, the chemical condensing agents cyanamide and dicyanamide may have assisted in the prebiotic synthesis of these membranogenic compounds.

In order to test the behavior of lipids under simulation conditions, Deamer and Gail Burchfield[8] allowed phospholipid liposomes to be subjected to the dehydration-hydration cycles in the presence of 6-carboxyfluorescein and salmon sperm DNA, respectively. During the hydration period the liposomes fused into multilamellar structures with the dye or the DNA trapped between the layers. When the system was rehydrated, the lamellae swelled and formed large vesicular structures containing the test substances. Encapsulation of prebiotic molecules was clearly a common feature of the cycle.

But it seems there was something more to the behavior of liposomes than containment that assisted the cell on its course. Biosynthesis is closely associated with structure, for it is the positioning of molecular species for reaction that is important in distinguishing cellular mechanics from open solutions. When the model system containing several phospholipids had d-AMP, d-TMP, and a template of DNA (poly dAdT alternating copolymer) added to it and was put through the "tide pool" process where the components were dried under anaerobic conditions for a few hours at 60–90°C, a phenomenal thing happened. A small amount of the monomer, either d-AMP or d-TMP, was incorporated into the polymeric molecule.[9]

The yield was small (0.1 percent per cycle) by laboratory standards, but immensely significant in implication. Both the lipid and the temperature were essential for the reaction. Deamer surmises that the lipid-dependent polymerization of nucleotide monomers into larger molecules may occur because the orientation of the monomers into the two-dimensional space offered by the lipid layers during the dry down cycle aligns them to form the phosphodiester bond.

Deamer and Hargreaves believe that the initial membranes on primordial earth were partly or largely lipid. Even before biosynthesis became a reality, there could have been selective forces favoring stability that concentrated available phospholipids into particular cells. The ionic and hydrophobic interactions between the membrane lipids and peptides would have allowed the latter to become intercalated in the membrane structure, perhaps injecting selective permeability. And eventually, the synthesis of proteins would have evolved proteins incorporated into the membrane for specific functions as enzymes or as agents for the transport of substances into and out of the cell.

All natural forces existed in the primordial environment for self-assembly of primitive cells. From inanimate substances to macromolecules and cellular structure, the bridge over the chasm to primal cells followed spontaneous processes. Considering the number of possible occurrences, the reaction series must have been funneled repeatedly into cellular species that crossed the threshold to biological evolution. From that moment on they were in the arena to win the future.

24
The Emergence of Cells

The primordial scene no longer exists. The prebiotic substances that gave birth to the first living cells are long gone, consumed and transformed in countless recycling processes by the very organisms they created. Nothing remains as witness to that event over three and a half eons ago on Archean Earth when matter and energy came together to create a living entity that eventually was to spread over every niche in the thin watery and gaseous film that sheaths the globe, and then reach with tentacles outward toward the immense universe.

To reconstruct what happened between three and four billion years ago, we are required to study the patterns of life processes, establish discernible trends of behavior, and attempt to retrace the steps that life has taken. Only then can we tread back into time, back beyond the beginning of the metazoans, the coelenterates, the sponges, beyond when eucaryotes appeared, back when oxygen was no longer a sustenance of life, back to the hot, strange world of the Archean, when shallow oceans skimmed around juvenile continents, scarred and blackened by the volcanic convulsions that gave them birth.

The prebiotic substances that formed on primordial earth washed into low-lying basins and collected. There they interacted with each other through a multitude of possible chemical combinations, some hydrolyzing quickly to recombine again another way; others more stable, accumulating. Life resulted from combinations that survived. But survival of chemical substances, normally regarded in the sense of resistance to decomposition, in this case was accomplished by a vastly superior assurance of survival. The manner in which particular combinations succeeded in perpetuating their existence was by making copies of themselves.

It began with the nucleotides. These chemicals could combine with each other in various weak associations through hydrogen bonding, some combinations stronger than others, but all dissociating in solution rather easily. When converted to activated derivatives such as polyphosphates, phosphoramidates, or cyclic 2',3'-phosphates, the nucleotides condensed into short chains. More often than not, the monomers became joined through 2',5'-linkages, as well as the 3',5'-bonding found in biological systems.

In order to carry out effective replication, the polynucleotides of an organism would have needed to be exclusively one type of linkage. By what means did the primeval cells have access to polynucleotides of a uniform bonding?

David Usher[1] of Cornell University has given a scheme whereby nucleotides containing both 2',5'- and 3',5'-linkages would gradually have become the normal 3',5'-polynucleotides from the effect of a cyclical pattern in nature. When the 3',5'-bonded oligonucleotides twist into a double helix, they have a greater resistance to hydrolysis than the 2',5'-bonded nucleotides that disrupt the helical configuration. By exposing mixed polymers to the conditions of a natural cycle, the preferential hydrolysis of the 2',5'-bond would favor the accumulation of the biological polynucleotides.

In Usher's scheme the nucleotides (nucleoside 2',3'-phosphates) would warm and bake in the presence of imidazole and other abiotic catalysts; this was followed by cooling and the addition of a small amount of water. When the cycle was repeated over and over, Usher suggests there would have been a progressive accumulation of oligonucleotides with the 3',5'-linkage. It is a process that would have occurred frequently on primordial earth.

Each morning as the sun rose high over the Archean landscape, the heat of the sun condensed the nucleotides into mixed oligomers. With the passing of the day, the temperature dropped, bringing the dew and dampening of the polymers. During the nocturnal interval, some 3',5'-bonded oligomers in solution found complementary partners to form short helices. As the sun rose again to repeat the cycle, those nucleotides wrapped in helices were slightly protected against hydrolysis, while the unprotected 2',5'-bound nucleotides were subjected to preferential hydrolysis. Day after day the cycle went on, each time more polynucleotide than before survived with the 3',5'-linkage in longer and longer chains.

The nucleotides with their purine and pyrimidine bases were mere chemical substances. Nevertheless, they had the ability to form hydrogen bonding, not with other nucleotides like themselves, but with a nucleotide of the other base type. In so doing, the polynucleotides were making chains of complementary nucleotides that formed the most stable associations. Only through separation of the double-stranded structure and the making of a copy of the "negative" did replication of molecules on earth begin. When this occurred, it became the basis of perpetuating molecular species.

For the polynucleotides to continue to form, they needed an applicable activation procedure and shielding from the intense ultraviolet rays that wreak destruction on the nucleotide structure. Sediments protected those polynucleotides that were covered but allowed little opportunity for further interactions. Only when they were carried into depths beneath a thick cover of water were the polynucleotides free to move in solution, while being spared from the fierce radiation.* Those replicating molecules that were washed into catchbasins eventually found themselves encased in vesicles formed from the elastic lipid film that covered the primordial waters.

In such an arrangement, they were held in close association, while having the advantage of movement that they had with the concentrated solution where there was a small amount of water. But to survive in water they needed to be supplied with free energy, which is to say, phosphorylation. It appears that it was at this stage that the nascent cells developed a rudimentary form of photosynthesis.

In his paper on the origin of photophosphorylation, Stillwell[2] suggests that, under the reducing conditions and intense ultraviolet light of primordial earth, the early forms of cells could have used photophosphorylation based on quinol phosphates to produce their ATP and other activated nucleotides. In contempoary cells quinones are present in the photophosphorylation chain, but chlorophyll and carotenoids are the initial collectors of visible light. Stillwell reasons that the series of reactants originally began with quinones and was extended to more effective absorbers of visible light as the ultraviolet rays lost their intensity when the ozone layer began to form.

---

* Ultraviolet light of 180 nanometers penetrates less than 1 centimeter, whereas the 280 nanometers wavelength penetrates 10 meters of water before reaching extinction.

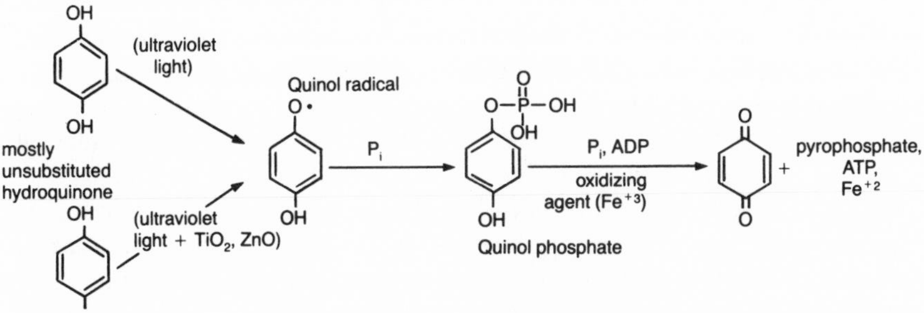

Figure 24.1. Proposed model for photophosphorylation using ultraviolet light to produce ATP.

Ultraviolet light was destructive in large amounts, but it was also the most plentiful energy source to be captured by simple substances. When hydroquinones are exposed to ultraviolet light in the presence of inorganic phosphates, quinol phosphates are formed. Quinol phosphates are strong phosphorylating agents, and together with a water-soluble iron sulfide complex (the forerunner of ferredoxin), they phosphorylate ADP and ATP. Presumably the other nucleotides were capable of being activated in the same manner.

The advantage of this model is that it shows that photophosphorylation could have developed early in the formation of life from simple substances without the complex membrane structuring that has been found with the later photosynthetic mechanism. As the genetic apparatus of organisms became larger and more susceptible to damage by ultraviolet light, the cells could thrive in depths or areas with optimal light for photophosphorylation and survival.

Eventually, when organisms evolved to the point of greater synthesis capability, they were able to develop photosynthesis based entirely on visible light activation using porphyrins, the class of chemicals that includes chlorophyll. Stillwell points out that even after porphyrins were being used with hydrogen sulfide, hydrogen, and reduced organic compounds as electron donors, phosphorylation may still have been through quinol phosphates. Only after the porphyrin-quinone-iron sulfide complex became lipophilic and attached to an elaborate membrane structure did quinones lose their role as phosphorylating agents and become only electron and proton shuttles in the photosynthesis mechanism, a role they currently perform.

These primitive cells with their rudimentary form of photophosphorylation and replicating polynucleotides could have been numerous, but life was on an extremely retarded pace. The reactions were slow—sometimes taking days, months, even years—for there were no enzymes. The lipid membranes were not generated by biosynthesis, but still relied upon the accumulated prebiotic lipid deposition. A crude form of exchange of material probably existed between vesicles when two coalesced, then redivided, or when an existing cell would come in contact with and absorb more lipid and separate into two different cells. During this period the cells were probably in confined environments such as volcanic lakes or catchment areas where encounters would be favored.

The primeval cells may have existed at this stage for millions or conceivably hundreds of millions of years. There were amino acids and other organic substances in the aquatic environment, but they would have been of no particular significance. Amino acids that passed into the cells may have, while still as the imines or by their amino groups being complexed on some inner cellular component, occasionally formed adenylates and subsequently, short peptides. But these probably had no consequences on the cells at the time except to represent clutter.

The polynucleotides would have been subjected to mutational change during this period, but the greatest variety could have resulted from the exchange of short polynucleotides between the population of cells. Coupling of these units into longer ones over a long time span would have produced a tremendously large number of combinations. It seems reasonable to assume that it was one of these polynucleotides that, after it was transcribed to a complementary chain, gave a molecule that folded back on itself and assumed a three-dimensional shape held together by opposing complementary bases. This polynucleotide became the first transfer RNA; the nucleotide chain from which it was transcribed, the first gene.

Because transcription is a one-on-one nucleotide relationship, the gene for the first transfer RNA needed to have been no longer than the transfer RNA itself. Margaret Dayhoff and her colleagues[3] at the Georgetown University Medical Center in Washington have studied the molecular evolution of the transfer RNAs and feel that the evidence strongly indicates that the transfer RNAs are all derived from a single gene. All transfer RNAs are synthesized from C, G, A, and U, each contain an identical tetramer sequence in one lobe, all end in CCA, and all have about the same length. They point out that the probability of even two such similar molecules occurring independently and simultaneously in the same cell would have been extremely small. On the other hand, duplication of the transfer RNA gene, followed by independent mutational changes in the separate genes, would have given closely similar products. Subsequent doublings expanded the number of transfer RNAs to accommodate a variety of amino acids.

The ancestral transfer RNA either accepted amino acids from amino acid adenylates or the terminal CCA end of the transfer RNA could have been activated and coupled with amino acids directly. In any event, the transfer RNA apparently began as a nonspecific intermediary in condensing amino acids when the tRNA-amino acid complex aligned on a polynucleotide chain serving like a contemporary messenger nucleic acid. In what manner transfer RNAs evolved their specificity is not yet understood, although the specificity linking each transfer RNA to a particular amino acid resides in the enzyme that promotes the attachment. But eventually the transfer RNAs managed to bring into operation the mechanism for condensing the amino acids into peptides correlating to the nucleotide sequence on the messenger molecule.

Ribosomes are subcellular particles that play an essential role in the synthesis of polypeptides. Without ribosomes the tRNA-amino acid complex does not remain attached to the messenger RNA long enough to react. For the primitive cell to begin the mechanism, it needed something to participate as a ribosome or to have carried out the step in some manner without the aid of these particles.

Francis Crick and others[4] have speculated that the initial protein synthesis may have occurred with transfer RNAs that bond to five base-pairs, instead of three, on the messenger RNA as with contemporary organisms. In this way the binding would have been sufficiently strong to hold the tRNA-amino acid complex in place long enough for the condensation to occur without the ribosome. They further proposed that, because of the resulting restrictions this placed on the number of codons, only four amino acids—glycine, serine, aspartic acid, and asparagine—were coded for in the original mechanism.

On the other hand, contemporary ribosomes are approximately one-half nucleic acid, with the remainder a variety of proteins. It seems more likely that protein synthesis began on primitive ribosomes formed from complexes of polynucleotides, and the triplet served as the codon from the beginning.

All amino acids used by later organisms may not have existed prebiologically. Aspartic acid, glycine, serine, and alanine, the four most common α-amino acids generated from cyanides and found prevalent in ferredoxin, most certainly were prebiotic. But arginine and histidine, which have not been produced in simultated experiments, presumably originated only through biosynthesis as organisms evolved.

There are two categories of amino acids: structural amino acids and amino acids with functional groups on the side-chain. Some of the latter, notably ones with an attached basic group, play vital roles in enzymes but are not readily formed by simulation experiments. As a result, other prebiotic compounds may have acted as substitutes until the biogenic synthesis of these amino acids developed. An example is 4-amino-imidazole-5-carboxamide, the imidazole derivative from ammonium cyanide, which may have served in the capacity of histidine for the emerging cell. Tryptophan, phenylalanine, tyrosine, and methionine are other possible candidates for a biosynthetic origin.

Once the primeval cell established a degree of specificity to transfer RNAs with a correlation of amino acids to nucleotide sequences in the messenger polynucleotide, any useful peptides would no longer be a chance happening but would be able to be produced as needed, giving the cell a selective advantage. One of the first peptides was the ancestral ferredoxin, whose origin Eck and Dayhoff[5] have traced back to a tetrapeptide.

Ferredoxin, now an iron sulfide protein of 55 amino acids in advanced forms of life, evolved to a progressively larger molecule because each change introduced a protein that was more efficient as an electron carrier than the predecessor.

There remains little doubt that this is the manner in which enzymes originated.

The primitive cell was probably already using iron sulfide as a catalyst, either alone, attached to cysteine, or with some abiotic peptide. The evolutionary development of ferredoxin seems to have begun with the peptide ala-asp-ser-gly. The peptide's gene, 12 nucleotides long, eventually doubled; and as the synthesizing abilities of organisms became more versatile and efficient, the genetic mechanism was able to incorporate other amino acids. Cysteine was among these amino acids, and the sulfide bond became attached to the iron sulfide. Mutations which modified and complicated the amino acid sequence created a series of changes that ultimately led to an intricate protein-iron sulfide complex of greatly enhanced efficiency.

Enzymes are not singular substances but seem so only because their extreme efficiency pales any other catalysts by comparison. This misconception has created the impression that no kind of biological system could exist without them. Certainly no organism comparable to contemporary life could, but for several hundred million years there may have been primitive cells existing before enzymes evolved. Once enzymes developed, however, their selective advantage to the cells was so immense that cells with enzymes spread rapidly, sweeping away all primitive life.

Enzymes evolved from simpler, less efficient substances that performed the same function they do. Histidine, pyridoxine, iron sulfide, and others are all substances that catalyze chemical reactions. They are also components that became part of or complexed with peptide chains in combinations that promoted the reactions faster than they alone. Eventually, enzymes evolved out of the developing protein synthesis that had these catalytic chemicals incorporated as a part of their structure to be the biocatalysts.

The first enzymes to have evolved would have been those that facilitated the propagation mechanism. Any advantage to reproduction sent the innovation spreading with the faster-breeding cells throughout the population. In this way, ferredoxin, by accelerating photophosphorylation, gave an advantage to the primitive cell. The structural detail of the enzymes associated with polymerization and translation processes is not yet elucidated to the point that the enzymes' origins can be defined, but presumably these enzymes would have been among the earliest to appear in biological systems.

The oldest fossils of organisms ever found on earth have been discovered in the Fig Tree and Onverwacht formations in the Barberton Mountain Land of the eastern Transvaal of South Africa. The fossils are in chert that occurs sometimes in horizontal layers but more often in pockets that had been in the surface of some ancient lava. There, in the mineral-rich waters of volcanic lakes, the earliest forms of life on earth thrived, died, and were buried in sediments to become entombed for three and a half billion years.

Were these the primitive cells that first formed from an assemblage of prebiotic substances to form life, or were these life after several hundred million years of perfecting the genetic and metabolic mechanisms? In a comparative study of the evolution of cytochromes and transfer RNAs, McLaughlin and Dayhoff[6] have computed the appearance of procaryotes to be 2.6 times more remote than that of eucaryotes. If the eucaryotes evolved 1.3 billion years ago, as suggested by Preston Cloud's microfossils, the primitive cells may have existed as long ago as 3.9 billion years.

This fact is quite remarkable. Considering that the earth was formed 4.6 billion years ago, followed by the *T. Tauri* stage of the sun, this allowed less than a billion years for the atmosphere and waters to form from volcanism, for the prebiological organic compounds to be produced and accumulated, and for the primitive cells to arise and evolve into fully functional organisms. These 0.7 billion years represented approximately the time span of the first orogeny, where the heat from radioactivity in the earth built up and went through the cycle of mountain building. So early did the primitive cells occur that the processes that led to their formation must have been highly probable. So probable, in fact, was the event of their appearance that it must be regarded as having been inevitable.

25
The Phenomenal Cell

The primeval cells that floated and rolled languidly in the warm waters of Archean volcanic lakes nearly 4 billion years ago were the beginning. They were genetically alive. They had crossed the threshold and no longer were merely inanimate molecules; they were an association of components that formed a unit in which nucleotides were activated, condensed to polynucleotides on other polynucleotides, and in the process performed molecular reproduction. But still, it may have ended there except for one critical feature—these earliest cells had the capacity for change. The door was left ajar.

The cells existed as a collection of chemical reactions, and any improvement in the efficiency of these reactions made the cell more competitive for the substances in the medium. The cells that became prevalent were those with access to components that were most effective in reproduction and accelerated rates of reaction. It was this ability to change, with the advantage being bestowed upon the most efficient, that carried the primitive cells from being genetically alive substances to organisms as we know them today with their remarkable powers of metabolism.

The path leading to metabolically alive cells meandered through thickets of change, each improving upon the cells' capacity to catalyze their biochemical reactions. These were to culminate eventually in the most impressive catalysts ever to exist—the enzymes. But before the enzymes were to become a part of cellular mechanics, there must have been a long interval during which the basis of their emergence was being laid.

The amino acids, purines, pyrimidines, and lipids entered the biological systems fully formed. These were the building blocks for the condensed polymeric components and cellular structure. The proteins, however, are composed of a greater variety of amino acids than seemingly would have been available in the reservoir of prebiotic substances. Nevertheless, few new amino acids could have been introduced after the beginning of protein synthesis. Once proteins became the mainstay of cellular function, any change in a single amino acid would have affected not one but a large number of proteins. For that reason, it appears that the amino acid series existed before the advent of the enzymes.

Amino acids are usually thought of in terms of subunits of proteins, but they also serve as precursors to other biochemicals, including other amino acids. Those amino acids produced most readily from cyanides, namely, alanine, serine, aspartic acid, and glycine, would have been in the environment in the greatest amounts. It is from these same few amino acids that contemporary cells synthesize a large variety of their basic components, presumably by chemical pathways passed down from the early primitive cells. Even before enzymes, then, the amino acids of biogenic origin must have been synthesized in primitive cells by simple chemical transformations that were the precursors of today's biochemical reactions.

The synthesis of the amino acids generally involved the transfer of functional groups— $-NH_4$, $-CH_3$, $-CO_3$, $-CH_2OH$, $-CHO$ —from one component or amino acid to another. By group transfer reactions a few amino acids can be converted to a wide variety of other amino acids and essential biological constituents. It is by these reactions and others that aspartic acid serves as the precursor to purines, pyrimidines, alanine, threonine, lysiene, methionine, and arginine. Serine, another amino acid from hydrogen cyanide, can degrade to pyruvate; the decarboxylation of pyruvic acid can, in turn, lead to acetate; and acetate in the activated form as acetyl coenzyme A is directly involved in the synthesis of lipids for the cellular membrane.

It is impressive how few substances can serve as precursors to other vital constituents. In some organisms, pyruvate, acetate, and carbonate—substances that would have been common on primordial earth—can furnish all the carbon to serine, glycine, cysteine, alanine, valine, leucine, isoleucine, lysine, aspartic acid, threonine, methionine, glutamic acid, proline, and arginine.[1]

All the chemical reactions used by biological systems already existed before life. The developing cells, therefore, did not create new chemical reactions to synthesize their constituents, but merely adopted means of exercising control over the many possible combinations. They accomplished this by accelerating select reactions with catalysts; and of the catalysts used that became permanently incorporated in the cellular metabolic machinery, some of the most important were the coenzymes.

Enzymes are proteins with catalytic properties. They can be simple polypeptide chains or they can be conjugated proteins with a nonprotein moiety attached for a specific function. A coenzyme, therefore, is the nonprotein portion of an enzyme whose presence is required for the activity. It may be regarded as the dis-

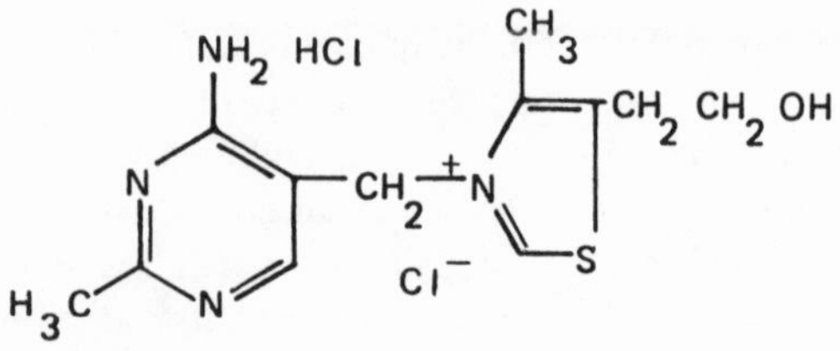

Figure 25.1. Thiamine.

sociable portion of conjugated proteins. Many coenzymes contain vitamins as part of their structure, and generally they function as acceptors of electrons or functional groups.

Like other fundamental components, most of the coenzymes appear to be of ancient origin, and they and their immediate antecedents probably became a part of biological systems fully formed. The pyridine nucleotides NAD and NADP became hydrogen acceptors for oxidation–reduction reactions, and coenzyme A became the instrument for acyl transfer. These are nucleotide derivatives of relatively simple composition that could have arisen abiotically. Thiamine, pyridoxal, and biotin were coenzymes used in group transfer.

Thiamine, as thiamine pyrophosphate, in an alkaline medium catalyzes the

$$\overset{O}{\underset{\parallel}{}}\quad\overset{O}{\underset{\parallel}{}}$$

cleavage of $-\text{C}-\text{C}-$ with the formation of an "active aldehyde." In this way, thiamine or its primordial counterpart accelerated the breakdown of pyruvate to acetate to be available in forming acetyl-S-Co A. Thus acetyl and other acyl groups were converted to the activated state for glyceride synthesis in order that the primitive cells could commence preparing their own lipid membrane.

Pyridoxal phosphate participates in several types of reactions. It is able to catalyze decarboxylations, deaminations, and transaminations of amino acids; and it also can transfer sulfur from methionine to serine to form cysteine.

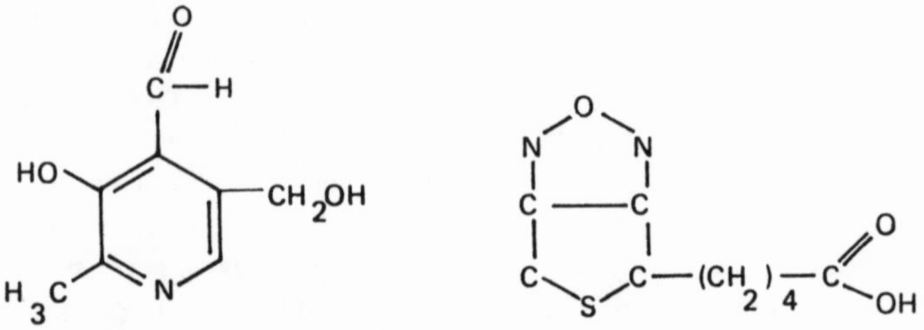

Figure 25.2. Pyridoxal and biotin.

Biotin functions as a coenzyme in carbon dioxide fixation and decarboxylation enzymes. A number of other enzyme reactions are reportedly affected by biotin.

These include succinic dehydrogenase and decarboxylase, as well as the deaminases of aspartic acid, serine, and threonine. Biotin is also involved in the synthesis of carbamyl phosphate, $H_2N - C - O - PO_3H_2$, a compound required in
$$\overset{\displaystyle \|}{O}$$
pyrimidine synthesis.

The biosynthesis of purines and pyrimidines, the bases of nucleic acids, is found in all organisms and probably was one of the earliest syntheses that was necessary for survival. In contemporary organisms, pyrimidines are synthesized by a scheme that must be original because of its universality and simplicity. In this procedure carbamyl phosphate reacts with aspartic acid to give ureidosuccinic acid, a compound that cyclizes to orotic acid, the precursor of the pyrimidines.

Both aspartic acid and glycine, as well as formate and carbon dioxide, are utilized for purine biosynthesis, while two of the nitrogen atoms of the purine come from glutamine.

Thus, within the primitive cells that began as being genetically alive, there was a sizable collection of spontaneous chemical reactions going on at varying rates that was eventually to create a metabolically active organism. From these reactions the primitive cells received an assortment of amino acids that became incorporated into the early biogenic peptides. And the cells that contained catalysts such as imidazole derivatives and coenzymes began generating other biochemicals that were serving as useful components. Most reactions were at a slow rate, and those primitive cells which lacked substances to accelerate useful transformations would eventually have been lost by being dissociated or absorbed by more successful cells.

The ascending road to survival was open to cells in which reactions were faster and ever more specifc, for efficiency was the path away from uncontrolled interactions and toward specific reactions that synthesized specific products. The coenzymes are general catalysts for a type of chemical reaction. When they became attached to polypeptide chains, however, their movement became restricted and closely associated with the peptide's interaction with the substrate. Rudimentary enzymes came into being. Slowly, step by step, as enzymes improved by amino acid substitution and enlargement, they narrowed the catalytic role of the coenzymes to reactions that are specific for only those molecules that can act as a substrate by fitting into the shape of the protein. By producing catalysts for specific reactions, the chaos of having many reactive species together in the cellular medium became organized into regulated reaction series. And as the control of the cellular reactions shifted to the synthesis of enzymes, biosynthesis became genetically programmed.

The synthesis of its own catalysts created one of life's most dynamic features. It became autocatalytic. Those cells that evolved enzymes that accelerated the synthesis of enzymes to accelerate the reactions ever faster quickly outpaced all other primitive cells. It was a cyclical process that essentially isolated the cellular mechanics from the environment and established the cell's independent nature. With an input of energy to drive its processes and feedback mechanisms for self-regulatory control, the cell was launched on its course into biological evolution. The use of cycles eventually became a dominant theme in the formation of self-sustaining life forms.[2]

Like all fundamental reactions of biological systems, the breakdown of glucose to lactic acid existed before enzymes. Ch. Degani and M. Halmann[3] have shown that the alkaline degradation of glucose 6-phosphate proceeds without enzyme action by parallel and consecutive reactions (see equation opposite).

The cell capitalized on the existing glycolysis by directing the series of spontaneous reactions through pyruvic acid and, with the assistance of the coenzyme NAD, tapped some of the chemical energy released in the degradation to produce two molecules of ATP.

This, then, became a means of extracting energy from the chemical structure of substances in the environment. Glucose was elevated to a central role as a

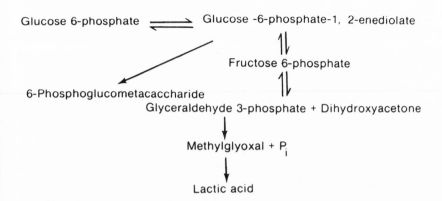

Glucose 6-phosphate ⇌ Glucose -6-phosphate-1, 2-enediolate

Fructose 6-phosphate

6-Phosphoglucometacaccharide

Glyceraldehyde 3-phosphate + Dihydroxyacetone

Methylglyoxal + P$_i$

Lactic acid

source of chemical energy when organisms completed the cycle by generating glucose from pyruvic acid. In addition to glucose, the clostridia, some of the most ancient anaerobes, were able to use alcohols, carboxylic acids, and amino acids for substrates in fermentation—substances that would have existed in the prebiotic reserve.[4]

It has been discovered that clostridia, as well as the photosynthetic bacteria, are capable of carbon dioxide fixation in a reaction driven by reduced ferredoxin.[5] The carbon dioxide assimilation carboxylates acetyl coenzyme A to pyruvate, which leads to the formation of the amino acids: aspartic and glutamic acids and alanine.

Until recently it was generally believed that the earth's aboriginal organisms were heterotrophs that extracted chemical energy through fermentation of the reserve of prebiological organic substances. But the archaebacteria, especially the methanogens, may be more ancient.[6] If the primordial atmosphere had been formed by the outgassing of volcanoes, as now believed, it would have consisted principally of carbon dioxide, water, nitrogen, and hydrogen. The reduction of carbon dioxide by hydrogen to methane and water is a spontaneous reaction, but the rate is slow enough that it could have been harnessed by a biological system. The methanogens apparently derived their chemical energy from this particular reaction. Some contemporary methanogens can use other sources that are convertible to carbon dioxide, such as formic acid, but none can metabolize typical organic sources like more advanced bacteria can.[7]

The primitive cells began as genetic replicating systems drawing their components from the prebiotic reserve. The number of amino acids was expanded through transformations, and eventually the synthesis of enzymes brought into bloom the metabolic processes of fully functional organisms. These were all extremely early adaptations which established life patterns for all forms of life on earth. Proteins have increased in number and composition ever since, but the amino acids became unchangeable, for any further change of a basic building block would have a devastating effect on the efficacy of thousands of proteins in an organism.

While the fundamental nature of life was being established, a peculiar thing was happening. Some cells began to use selectively more of one optical isomer of amino acids and ribose than another. The differentiation probably began slowly but accelerated exponentially as cells consumed the debris of other cells, until within a relatively short time all cells were using exclusively L-α-amino acids and D-sugars.

Compounds with an asymmetrical carbon (carbon with four different substituents) have two stereoisomers for each of these carbons. The isomers are non-superimposable mirror images of each other, and as a consequence, one form rotates polarized light to the left (−), the other rotates it to the right (+). The configurations of the optical isomers are designated D and L to indicate not the direction of rotation but an arbitrary assignment of the substituents at the α-carbon of amino acids, the 4-carbon of pentoses, such as ribose, and the 5-carbon of hexoses, such as glucose.

The stereoselectivity by the primitive cells was a giant step in the direction of organizing matter into the functional system we call life. Since the free energies of optical isomers in an optically inactive environment are identical, the synthesis of amino acids, sugars, and other asymmetrical substances, except by organisms, gives an equal number of the possible stereoisomers. Such racemic mixtures fail to exhibit any optical activity. So unusual, then, is this property of biological molecules that it has been regarded as a characteristic of living systems and has been a principal property tested in organic substances of meteorites, geological deposits, and planetary explorations to determine whether or not they are of biogenic origin.

Ever since 1815, when Jean Baptiste Biot discovered that sugars, tartaric acid, camphor, and turpentine oil rotated plane polarized light in the liquid state or solution, the riddle of the origin of optical activity has intrigued scientists. Pasteur learned that it was a feature intrinsic to the asymmetry of individual molecules. Thinking the earth's magnetic field might be the dissymmetrical factor that induces asymmetric synthesis, he tried unsuccessfully to prepare optical isomers by using powerful magnetic fields. One reason for his lack of success was that his premise was wrong. The magnetic field and rotation of the earth are actually symmetrical forces.

In the laboratory optical isomers can be separated only with great difficulty, and then usually with the assistance of another purified isomer. For years stereospecific synthesis remained an elusive goal. So frustrated were chemists in their attempts to synthesize stereoisomers that F. R. Japp[8] in 1898 resurrected Vitalism by declaring that the primary synthesis of asymmetrical molecules was impossible and that, like the living organism, asymmetrical molecules could be derived only from asymmetrical molecules.

Japp was not completely correct in his conclusion, but he did emphasize the question: How could the optical activity of biological systems have originated if only living cells are able to produce it?

Theories to explain how biological systems became stereohomogeneous fall into two categories: the before life theory and the after the origin of life theory. The before life advocates try to find a physical means that could have produced stereospecific synthesis of the prebiotic compounds; the others contend that stereohomogeneity is of biological origin.

Van't Hoff[9] in 1894 considered polarized light as a possible factor that could have caused asymmetrical synthesis under natural conditions. It has since been suggested that partially plane polarized light from the sky, converted into partially right-handed circularly polarized light when reflected on the earth's surface under the influence of the magnetic field, could have resolved optical isomers. This hypothesis has been tested experimentally by the stereoselective photodestruction of isomers with circularized polarized light.[10] The percent resolution achieved, however, was extremely small, even after half of the starting material had been decomposed. The significance of the results was further diminished by the fact that natural light is not as polarized as that used in the experiments. As an explanation of the origin of stereoisomerism, the tests were unconvincing.

There are many optically active crystals in nature, an example being the right- and left-handed crystals of quartz. Upon melting or dissolution, quartz crystals lose their activity. The loss occurs because their optical activity is due not to asymmetrical molecules but to the helical arrangement of the -Si-O-Si-O- chains in the crystal. Harada[11] used optically active quartz powder to orient the formation of amino acid crystals, but the activities were so small as to be near the limit of experimental error.

Under carefully controlled conditions it is possible to induce preferential crystallization of optical isomers from supersaturated solutions of racemic mixtures through seeding or spontaneously. Tartaric acid, malic acid, lactic acid, threonine, histidine, glutamic acid, and many others have been resolved by this procedure. The seeding technique is used in this way to prepare monosodium L-glutamate industrially in Japan. As appealing as the technique may appear, it is difficult to envision this as the origin of biological optical activity. It does not exhibit a selective preference of one isomer over the other, and the laboratory conditions are too idealized for a natural setting.

It has been postulated that natural radioactivity may have been a contributing factor in the stereoselectivity of early organisms.[12] M. Goldhaber and others[13] showed that beta-decay electrons emitted from ^{60}Co are polarized by spinning in one direction. Conceivably the dissymmetry of these elementary particles could affect molecules with an asymmetrical physical force. In a recent article in *Origins of Life*, H. P. Noyes, W. A. Bonner, and J. A. Tomlin[14] of Stanford University suggested that beta-decay electrons from ^{14}C in amino acids could have exhibited some selective destruction of the D-isomer. The reaction, however, appears to work only on leucine, and it is questionable whether the rate of conversion would be fast enough to exceed racemization.

Optical isomerism in amino acids and sugars is not an absolutely stable configuration. Optical isomers racemize and lose their activity by equilibrating toward an equal number of both isomers; optically active amino acids racemize through ionization of the α-hydrogen by a rate that is independent of pH between 3 and 8, but is affected by temperature. The respective half-lives of isoleucine, alanine, and phenylalanine are 35,000, 11,000, and 2,000 years at 25°C.[15]

The difficulty with the attempts to find a prelife explanation of optical activity is that all the conditions believed to have been present in the primordial environment favor racemization and not the accumulation of optical isomers. Stereoselectivity must therefore have been a procedure adopted by early organisms in some manner, and because of an associated advantage, spread throughout all life on earth.

One does not have to look far in molecular biology to see the answer. A replicating double-stranded nucleic acid can be constructed of either all D-ribotides or all L-ribotides, but it is not possible to build a regular structure with a mixture of the two.[16] To change from D- to L-β-uridine requires inversion of each substituent in the plane of the ribose ring. If an attempt were made to insert L-β-uridine into a double-stranded nucleic acid helix, the 3'- and 5'- hydroxyls would not join up correctly with the adjacent nucleotides.

Activated D-adenylic acid will condense with D-adenosine but not with L-adenosine, on a poly-D-uridylic acid template. The polynucleotides of the primitive organisms were optically active not because of some intriguing natural formation of particular isomers but because only stereospecific polynucleotides would have worked in the genetic mechanism.

Similarly, it is the helical configuration that gives protein its specific shape. And it is upon the specific shape that solubility, the capacity to crystallize, and the ability to enter into specific chemical reactions with other molecules depend. Limited helical conformation is possible with amino acids that are not optically pure, but the structure is severely hindered by steric effects. A random assortment of L- and D-amino acid residues, however, seems to make the α-helix impossible. The incorporation of only L-amino acids by the emerging cells probably occurred when peptide synthesis took place by translation of the stereohomogeneous polynucleotides and only amino acids of the same spatial orientation would align properly for the condensing.*

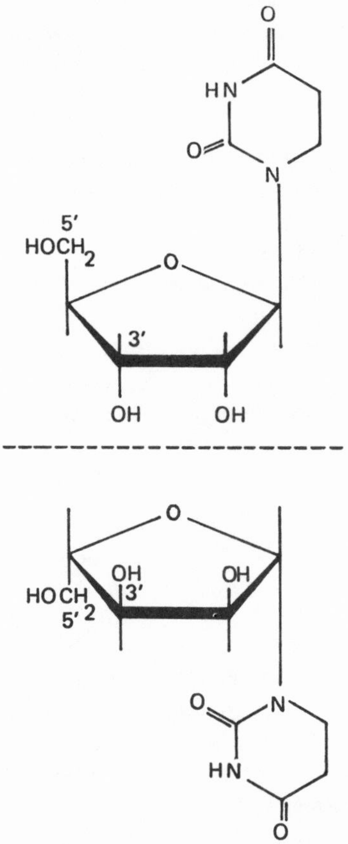

D–β–Uridine

L–β–Uridine

All biological systems on earth use the same optical isomers. Presumably, an organism using exclusively L-ribose and D-amino acids would function equally effectively as their counterpart that evolved. If the prebiotic substances were racemic, why did not both types of organisms arise? We can only surmise that

* There are exceptions to the stereohomogeneity of natural products. Antibiotics gramicidin, tyrocidine, and others are peptides with D-amino acids included in their chains.

the primitive cells that used D-ribose occurred first and were efficient enough to be synthesizing many of their components before any of their opposites appeared.

There have been attempts to dramatize the emergence of the first cell as a "war" between optical antipodes for the horde of prebiotic substances. This seems odd considering that neither would be using the isomer of the other. Rather, it seems more plausible that one form of primitive cell developed and the amino acids and sugars synthesized by its optically active enzymes quickly overwhelmed all biological substances with its form of optical activity.

If, therefore, essentially all organic matter on earth is of biogenic origin, the amount of prebiotic organic matter that existed on primordial earth has probably been greatly exaggerated. Harold Blum[17] has pointed out that if all the oxygen on earth corresponds to the amount of carbon dioxide that was assimilated into carbohydrates, this amounts to 4×10^{19} moles. At the present rate of metabolism, without being replenished through further photosynthesis, the organic material would last as a food supply only about 3,000 years. Since the present rate of use is considerably faster than in primeval times, a rough estimate of how long primitive life could have developed before evolving photosynthesis could not have been more than a few tens of thousands of years.

This estimate suggests that photosynthesis appeared very near the beginning of life. But it also indicates that there was no vast accumulation of prebiotic organic matter on primordial earth to sustain life for any length of time. There was no mythical primordial soup, except on a localized scale. Life began in a fairly sparse accumulation of precursors and survived only by becoming self-sustaining.

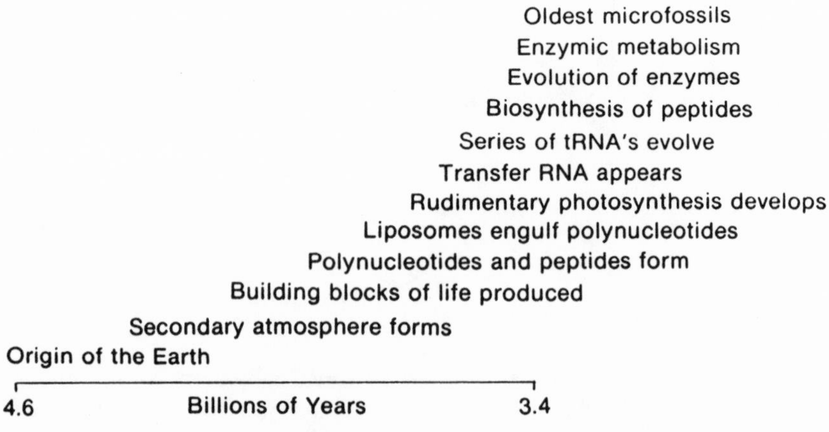

Oldest microfossils

Enzymic metabolism

Evolution of enzymes

Biosynthesis of peptides

Series of tRNA's evolve

Transfer RNA appears

Rudimentary photosynthesis develops

Liposomes engulf polynucleotides

Polynucleotides and peptides form

Building blocks of life produced

Secondary atmosphere forms

Origin of the Earth

4.6   Billions of Years   3.4

Figure 25.3. Sequence of events postulated to have led to the formation of a biological cell. Only the origin of the earth and the age of the oldest microfossil have been dated experimentally.

26
Other Ways, Other Places

The greatest obstacle to reconstructing how life began has been the success of its evolution. The efficiency and complexity of any living thing, even the simplest microorganism, is so awe-inspiring that it has created the impression that life could not have existed except as it exists now. The image of thousands of gigantic molecules orchestrated in precise harmony in packet too small to be seen evolving unaided from simple substances and becoming life challenges the imagination. Yet, because of the enormous length of time that life has existed on earth, evolution required progressive changes to occur in individual proteins no more often than every few million years. One difficulty is that our impressions are locked into our world of size and time. A lifetime, the age of civilization, the interval since man has been a species are but moments when compared to the 3 billion years it took the primitive cell to evolve to the level of the the jellyfish. For 80 percent of the time life has been on earth, it existed solely as single-celled microorganisms.

Locked in step with the earth's own evolution, life passes only once. The prebiotic substances that led to life's beginning could have formed only under the reducing conditions of the primordial atmosphere; eucaryotes arose only after a long evolution of procaryotes and in response to the oxygenation of the environment; multicellular organisms were possible only after the Pasteur point of free oxygen was reached; and colonization of the continents came after the ozone layer developed as a protective screen to the ultraviolet radiation.

Once it had come into being, life could go only in one direction—from the simple to the complex.* The forces of

* Viruses and rickettsiae seem to have evolved contrary to this principle, but they apparently are degenerate forms, instead of more evolved types.

Figure 26.1. Martian sunset over Chryse Planitia taken on August 20, 1976, by Viking 1. The camera began scanning the scene about 4 minutes after the sun had dipped below the horizon. The sun had set nearly 3 degrees below the horizon by the time this computer enhanced picture was taken.

selection favoring perpetuation of those variations that can best assure their survival and reproduction carried life forward. The result of that selection had been toward greater efficiency and complexity. As organisms passed beyond the primitive cell stage of complexity, their assembly ceased being intrinsic in their chemical structure and became interlocked with specific components whose syntheses were genetically programmed into the cells' composition. When organisms reached this stage there was no turning back—they could only go forward. Life had become committed to the treadmill where a cell can arise only from a preexisting cell.

As organisms attained the level of complexity where self-assembly was no longer possible, the specter of death became a part of life. The growth of complexity required perfect union. With each step of evolution, whenever the assembly of cellular constituents became dissociated, the chances of it coming together again into a functional unit became ever more improbable. Life lost became life irrecoverable. It was to be preserved only by passing it on to new generations.

The building blocks of biological systems are the nucleotides, amino acids, carbohydrates, and lipids; the architecture consists of the chainlike polymers of these units fitted together into a concerted cellular entity. The blocks have remained the same; it is the polymers that have grown and varied, molded by mutational change and the pressures of natural selection. The enzymes, whose nature determines the complete physical and chemical character of an organism, have continued to be subjected to evolutionary change, marching to the principle of continuity, each step in evolution following the preceding step in an uninterrupted advance for nearly 4 billion years.

The result has been that organisms have retained their fundamental nature but have divided and evolved into immense varieties to capture all available chemical and energy sources. The incredibly diverse and bizarre morphology of the insects and microscopic life attests to the many forms life can assume to this end. And the conditions under which organisms can survive to eke out an existence in an ecological niche is astonishing. The alga *Cyandium calidarium* can grow in concentrated solutions of hot sulfuric acid; sulfate-producing bacteria are reported to grow and reproduce at 104°C (219°F) under high pressures; and many organisms use organic and inorganic antifreezes to lower the freezing point of their internal liquids so that they can live at several tens of degrees below 0°C. Some insects use dimethyl sulfoxide for antifreeze.

The environments, therefore, under which life can exist are extreme and varied. The many examples illustrate the adaptability of the biological system through selection from variation in the enzymes and the processes derived from them. But still, all different forms of life are basically the same. They use nucleic acids for their reservoir of information and enzymes to catalyze their chemical conversions. The building blocks are identical; organisms differ only in the structures assembled from them.

In effect, when it comes to examples of life, we have only one.

The question is whether there is only one example of life because it can exist in only one form and therefore is unique, or whether it is simply the surviving one of a number of possible ways it could have occurred. We have already entertained the idea that organisms use L-α-amino acids and D-sugars, but there appears to be no reason why we could not have become composed of D-α-amino acids and L-sugars. Also, the genetic code could probably have been other combinations without any sacrifice of efficacy. There are cases where the selection occurred not because of some advantage, but because only one antipode and one code could be used, and the incidental circumstances at the time dictated the choice. Once living cells assembled, however, all living things from that time on became committed to one type of biological system for the remainder of life's existence.

Could the primitive cells have begun differently? Mary Ellen Jones of Brandeis University and Fritz Lipmann[1] of the Rockefeller Institute have suggested that early living organisms may have used inorganic polyphosphate instead of ATP for energy transformation. If pyrophosphate preceded ATP as the early energy-rich phosphate compound, it is possible that there are metabolic remnants of this among the primitive types of microorganisms alive today. This was, indeed, found to be the case. I. S. Kulaev[2] discovered that there are bacteria and fungi containing enzymes that catalyze the synthesis of inorganic polyphosphates instead of ATP, and Baltscheffsky[3] found eight reactions involving inorganic pyrophosphates in chromatophores of *Rhodospirillum rubrum*, a photosynthetic bacterium. If pyrophosphate had been used before ATP in cellular metabolism, it was probably superseded because of the nucleotide's ability to make finer and more specific contacts with cellular metabolites, as well as ATP's capacity to perform other functions that are inaccessible to inorganic polyphosphates.

There is also a consideration that early peptides may have been synthesized by a procedure that preceded ribosomal protein synthesis where the amino acid sequence is coded in the base sequence of a nucleic acid. Glutathione* and poly-

* γ-L-Glutamyl-L-cysteinylglycine.

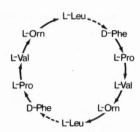

Figure 26.2. The structure of gramicidin S.

glutamic acid are uncoded peptides,[4] as are gramicidin S and tyrocidine, the microbial antibiotics.

Gramicidin S is a cyclic decapeptide containing a repeated chain of five different amino acids, including D-phenylalanine and ornithine, neither of which is found in proteins. Lipmann and his coworkers[5] found that two combined enzyme fractions from *Bacillus brevis,* when supplied with ATP, Mg^{++}, and the five amino acids, synthesized the decapeptide.

Apparently the pentapeptide is formed by a mechanism analogous to the manner in which fatty acids are synthesized. To synthesize fatty acids, the acetyl-S-coenzyme A complexes are condensed and reduced, while remaining bound to a multienzyme system. Lipmann's findings are that the pentapeptide of gramicidin is synthesized in a similar manner, beginning with D-phenylalanine and followed by the sequential addition of proline, valine, ornithine, and leucine. The synthesis of gramicidin is completed by the coupling of two chains into the cyclic decapeptide molecule. Bacteria produce tyrocidine in the same manner.[6] The length of polypeptides synthesized by this method is limited, the longest being antibiotics that are straight chains of 15 amino acids.

Fatty acid synthesis, being essential for the cellular membrane, must have been one of the earliest biosyntheses of the primitive cells. Lipmann[7] suggests that the similarity between the synthesis of fatty acids and that of the antibiotics could indicate that this procedure is more ancient than the ribosome-linked protein synthesis.

The sulfhydryl enzymes, however, are a common group and there is no indication that the enzyme specificity in attaching amino acids to small peptides is any different from the coupling of a chemical group in the synthesis of other compounds. The reaction is mediated by complex protein catalysts, and if it occurred before the development of enzymes, it would have been nonspecific. More probably this nonribosomal synthesis does not antedate enzymes, but bacteria evolved it because it imparted some advantage by having the antibiotics synthesized *in situ* in the cell wall or a similar location by enzymes.

Extending back a step earlier, we may ask if life could have assembled from building blocks other than the nucleotides and amino acids.

In order for the earliest cells to arise, they needed primarily a mode of mo-

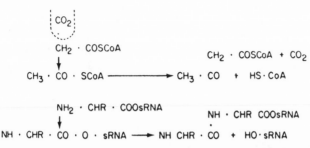

Figure 26.3. Comparison of fatty acid synthesis with noncoded peptide synthesis.

lecular replication to be genetically alive. The basis had to be some chemical structure that would adhere to other chemical units in a uniform way while they condensed, thus creating a copy. The bonding had to be strong enough to be functional, yet weak enough for the copies to separate after replication. And too, for the genetic material to be the informational center for all the cellular components in a metabolically alive organism, it had to have many units, which is to say, many information bits, within a large molecule.

The basis of the genetic material of life on earth is the purines and pyrimidines with their hydrogen bonding. These chemicals alone fail to fulfill all the requirements because they cannot condense with one another to make polymers. They can, however, condense with sugars, the sugars can be phosphorylated, and the nucleotide units can join through diester linkages. Interestingly, even after this involved process to make polymers containing the heterocyclic bases, the polynucleotides do not make copies directly, but rather, make "negatives" that act as the template for the copy.

It seems like a complicated procedure, but it survived and was adopted by the primitive cells for one reason: the structure of a polynucleotide could not only be transcribed into another polynucleotide, but also translated into chemicals of a different nature and different function—the peptides—which developed into powerful catalysts.

At first glance, amino acids seem to be the logical choice for forming informational molecules. They are varied and have a carboxyl and amino group that allow the units to link with each other. Despite being able to form polymers more directly than purines and pyrimidines, however, the amino acids fail as informational molecules because the information cannot be retrieved. The amino acid side-chains do not form precise interactions in the manner that characterizes the bases, and as a result, transcribing and translating polypeptides never became feasible for biological systems.*

---

* The agent responsible for scrapie, a disease afflicting sheep, appears to be a virus-sized replicating form that contains a protein but no nucleic acid.[8] Its mode of reproduction is speculative, but one suggestion, by J. S. Griffith,[9] is that the protein acts as an antigen to stimulate the production of an antibody that is identical to itself.

Life arose with its particular biological nature of purines, pyrimidines, and amino acids because these were the chemical substances available on prebiotic earth. Energy acting on the volcanic gases of primordial earth generated cyanides, and ammonium cyanide condensed principally to the purines and a select number of α-amino acids. The pyrimidines apparently formed less readily but would have been selectively absorbed to become incorporated into functional systems. What is most impressive is the relatively small number of chemicals formed from condensation, and yet these were sufficient bases to assemble cells that became self-sustaining forms of life.

Only four elements—carbon, hydrogen, nitrogen, and oxygen—make up 98 percent of the composition of protoplasm. Except for helium, these are the four most abundant elements in the universe. The occurrence of life cannot, therefore, be attributed to some rare element. Nor are the building blocks unusual chemical compounds, for they form readily from the condensation of cyanide. Of the 36 elements used by organisms, J. H. McClendon[10] found nine—C, H, N, O, P, S, Mg, K, and Fe—to be essential. The remaining 27 elements appear to have become incorporated not because they are essential for life but because they were useful and were available in the environment in sufficient quantity.

Of the nine essential elements, only C, H, N, and O are so unique as to be indispensable for the structuring of a biological system. The other five—P, S. Mg, K, and Fe—are elements whose chemical properties under conditions of earth make them essential for our developed biochemistry. These properties, however, are not restricted exclusively to these elements but are exhibited also to some extent in closely related elements. Conceivably, under different chemical environments, a biological system could evolve where arsenic substitutes for phosphorus, selenium for sulfur, manganese for magnesium, rubidium for potassium, and cobalt, nickel, or vanadium for iron.

Carbon, however, is the backbone of life. With its remarkable capacity for linking with many elements, including other carbon atoms, into an enormous variety of combinations, this element above all stands responsible for the architecture of biological systems. It is so indispensable and common that one readily assumes that any life form must be based on carbon. And yet there is another element, also common on earth, that can link itself and other elements into long chains and different combinations much like carbon. That element is silicon.

Silicon can form Si-Si chain compounds, but their stability is less than that of their carbon counterparts. The carbon-carbon bond is about twice as strong thermodynamically as the silicon-silicon bond. On the other hand, silicon has a strong affinity for oxygen, and the highly stable Si-O-Si-O chains form the crystalline arrays of quartz and many other minerals. Carbon also combines with oxygen readily, but the result is quite different. Whereas silicon dioxide is the component of sand, carbon dioxide is a gas; and it is the volatility of carbon compounds that led to carbon being concentrated in the earth's hydrosphere and atmosphere.

Silicon may never be a proper basis for a biological system, but there is a hypothesis that silicon minerals could have played a role in the origin of life. A. G. Cairns-Smith[11] of the University of Glasgow believes that a functional cell would have had to have been too complex to assemble from the prebiotic building blocks within a reasonable time and under ordinary circumstances. Instead, he proposes that life evolved through natural selection from inorganic crystals.

When crystals form, imperfections often appear in the lattice, and they are replicated in further crystallization. Since the replication of these imperfections is self-selective, it represents reproduction not unlike that of nucleic acids. In fact, many kinds of crystals could, in principle, hold a large amount of information, and the information density of a colloidal clay crystalite could be comparable to that of DNA.

Cairns-Smith contends that a crystal as a primitive gene could have controlled the development of organic macromolecules through the adsorption of organic molecules to it. Evolution of the primitive geneographs would have proceeded through selective elaboration that had survival value for the clay crystalites that held them. Eventually, the first organism is believed to have emerged by a genetic metamorphosis as the organic macromolecules gradually took over the control mechanism and the crystalites were discarded.

Although the chemicals from which life arose are common, the circumstances of the earth's evolution may be more extraordinary. The earth's size and distance from the sun were critical, for it was the heat generated from the disintegration of radioactive elements, unable to be dissipated into space, that carried the earth through a thermal evolution, liberating volatiles from the interior and creating the hydrosphere and atmosphere. Of all the planets, asteroids, and satellites in our solar system, the earth alone has liquid water in abundance.

Venus, the planet closest to the earth in size and space, became trapped in a runaway greenhouse effect and lost its water and all opportunity to Cytherean life as its surface temperature soared. Mars, with evidence of past volcanism, appears to be a planet in an ice age, with water that once flowed on its surface remaining sealed in polar ice caps. And beyond Mars and the asteroids revolve the planetary giants: Jupiter, Saturn, Uranus, and Neptune. Although these four outer planets are much larger, they differ from the terrestrial planets more significantly in chemical composition. They are composed almost entirely of gases. Jupiter is 80 percent hydrogen,[12] Saturn, 60 percent, with the balance almost certainly helium; and although Uranus and Neptune contain quantities of carbon, nitrogen, and oxygen, they too are globes of partly frozen, partly liquified, gases.

It is difficult to conceive of biological systems developing in the absence of water. Nevertheless, there may be parts of the solar system containing an aqueous environment with little resemblance to the earth that may have fostered the early stages of life formation. The atmosphere of Jupiter consists of methane, ammonia, and hydrogen—the primordial gases that can lead to prebiotic substances. This atmosphere, however, is not uniformly distributed but is stratified through a temperature gradient. Calculations by Gallet[13] suggest that below clouds of frozen ammonia crystals there should be a region of ammonia rainstorms; under the ammonia droplets, gaseous ammonia forms a layer; and a few kilometers beneath the ammonia clouds lie successive tiers of ice, liquid water, and water vapor. Carl Sagan[14] suggested that organic compounds derived from the ammonia and methane could interact in this zone to yield the larger molecules necessary for life.

To test this hypothesis, Fritz Woeller and Cyril Ponnamperuma[15] at the Ames Research Center carried out reactions on simulated Jovian atmospheres with an electric discharge. The volatile fraction from the experiment did, in fact, contain precursors of many biological compounds. But in addition to the volatiles, the reaction also yielded a nonvolatile fraction consisting of orange-red polymers. Woeller and Ponnamperuma speculated that this polymeric substance may be responsible for the color of the Great Red Spot of Jupiter.

The majority of large worlds in our solar system are not planets—they are satellites. Thirty moons orbit Jupiter, Saturn, Uranus, and Neptune, some of them extremely large.* Three of Jupiter's four Galilean satellites, Europa, Ganymede, and Calisto, are approximately the size of the planet Mercury, as are the Saturnian satellite Titan and Neptune's Triton. These are not hunks of debris like Phobos and Deimos that circle Mars; these are large bodies the size of our own moon and larger, each a different kind of world, some with thin atmospheres, some ice-covered. In this far region of the solar system there are entire moons with the density of ice, hundreds of miles across, circling planets like enormous snowballs in space.

The larger moons have low densities indicative of composites of rocky and icy material. It is speculated that within these large bodies the heat from radioactive decay produced melting, causing the silicates to settle to the core as compacted mud and forming a thin crust of ice that floats on a thick mantle of ammonia and water. Some surfaces seem to consist exclusively of ice, while others may be mixtures of ice hydrates of ammonia, methane, and various substances. These moons represent a whole new class of worlds with different compositions and structures, and they promise a whole new class of phenomena.

Titan, the largest moon of Saturn, appears to have a relatively dense atmosphere—several times denser than that of Mars—with red clouds that usually obscure the satellite's surface. Methane has been detected, ammonia is probably present, and it appears that there is an appreciable amount of hydrogen. The effect of the hydrogen is that it creates a greenhouse effect that elevates the temperature to approximately that of earth. If Mars proves to be lifeless, the next place within our family of planets that may have spawned life lies 1 billion miles out on a Saturnian moon called Titan.

We have deeply ingrained impressions of what life is and should be—all from

---

* The flights of the Voyagers have led to the discovery of many more.

Figure 26.4. Saturn seen from its largest moon, Titan, 760,000 miles away, as envisioned by the artist Chesley Bonestell.

our one example. As we explore other worlds we are probably opening ourselves to many surprises. Life arose on earth and assumed its particular nature because it was possible from the prebiotic substances and there was sufficient time for them to organize into a biological system. With enough time, what was possible became inevitable. Time is the parameter that allows it to happen and is analogous to a computer that scans all possible combinations until the solution is found.

Nevertheless, with other substances and other chemical environments, the chances of radically different biological systems evolving remains a possibility. Liquid ammonia may be a more common solvent in our solar system and through-out the universe than liquid water. If all the possibilities of life forming in the universe were explored, its occurrence in liquid ammonia might be prevalent.

Whereas water is liquid between 0° and 100°C, ammonia is liquid between −78° and −33°C.* Water has a high surface tension that makes spontaneous formation of cells with lipid–water interfaces ideal. Liquid ammonia is much less viscous, and polar substances are more soluble in water than in ammonia. Chlorides, sulfates, hydroxides, and oxides, for example, are insoluble in liquid ammonia.

But since our single known example of life evolved as a biological system adapted to an aqueous environment, the properties of water should naturally appear to us as the most ideal. A form of fish swimming in its sea of liquid

* At a pressure of 1 atmosphere.

ammonia would doubtless find our world intolerably hot and the chemical properties of water just as alien and distasteful.

Our life kind arose from prebiotic compounds that formed out of the gases from volcanic activity. If other worlds of different composition underwent volcanism, their atmosphere could differ significantly from that of primordial earth. Would they then have evolved a different biological system? The answer is probably no. The Miller experiment does not depend so much on the chemical composition of the gases as on the elements present. A reduced atmosphere containing carbon, hydrogen, nitrogen, and oxygen, regardless of the chemical form, when exposed to high energy will produce the same building blocks of life that formed on primordial earth.

Life probably occurs throughout the universe in many different forms and in environmental extremes, but it seems that the basis for it is likely to be the same everywhere. In this respect it is unique. Life is just something that happens. It is like the thermonuclear reactions in stellar evolution. It is a relationship between matter and energy that occurs spontaneously when the conditions are within narrow limits, and it evolves through stages. There are many varieties of matter and energy, but their fundamental chemistry and physics follow the same pattern—and that pattern is universal.

27
Organic Compounds in the Universe

One hundred and fifty years ago Friedrich Wöhler astounded the scientific community by heating ammonium cyanate and producing urea. In a simple experiment he showed that there was no insurmountable barrier between biological substances and inorganic compounds. In effect, Wöhler and Miller demonstrated the same principle: there are interconvertible relationships between the simple molecules of biology and the chemical nature of geology. It is a curious coincidence that Wöhler also studied a phenomenon of his day that remained unresolved until it was reinvestigated the same year that Miller carried out his experiment.

In 1953, while Miller was conducting his electric discharge experiment at the University of Chicago, a geologist in London inquired of the British Museum if he could analyze a sample from their meteorite collection. It had been known since the time of Berzelius and Wöhler that certain meteorites contained organic matter. But since organic material comes almost invariably from living things, the thought that it could be present in meteorites led to implications too astonishing to believe. From the very beginning, however, meteorites have been the sort of thing that invites credibility problems.

Since at least the dawn of history there had been folklore of stones, and even masses of iron, occasionally falling on the earth. But the Age of Reason brought a skepticism that regarded this as impossible. When a stone fell at Lucé, France, in 1768, a commission led by the eminent French chemist Antoine Lavoisier made an inquiry. The conclusion of the commission was that the meteorite was a terrestrial rock. Three more falls occurred in Europe in the 1790s in which specimens were preserved, but scientific opinion re-

Glon dem donnerstein gefallé im rcij.tar: vor Ensisheim

Figure 27.1. The fall of the Ensisheim meteorite, Alsace, 1492.

mained adamant. Only after a spectacular fall of thousands of stones at l'Aigle, France, on April 26, 1803, did the Royal Academy of Science of Paris concede that stones can indeed fall from the sky.

Meteorites fall into three major categories: irons, stones, and stony irons. Each of these is further subdivided by composition. Based on elemental and mineral content, there is a marked uniformity within the classes, with a great divergence between classes. The principal mineral constituents all occur as terrestrial minerals and the abundance of the elements in meteorites follows the regularities known for earth.

Because stony meteorites tend to weather quickly and are hard to distinguish from ordinary rocks, it is usually the irons with their damascened exterior that are seen in museums. But of meteorites recovered soon after a fall, 94 percent have been found to be stones. Of the stones, 80 percent are chondrites, characterized by rounded granules of enstatite or chrysolite embedded in the mass of meteoritic stone—and among the stony meteorites there is a small group called carbonaceous chondrites. Despite the fact that at least 2 percent of all meteorites are of this group, being friable and disintegrating rapidly by weathering, they are hard to recognize unless found soon after the fall. These meteorites received their name because some contain as much as 5 percent organic carbon.

There are 21 known carbonaceous chondrites, all of which were collected soon after the fall was observed. Berzelius[1] and Wöhler[2] both attempted to analyze the organic matter in carbonaceous meteorites, but the carbon appeared to be in the form of an insoluble polymer. Without identifying any specific compounds, they concluded that the organic substance was not of biological origin.

Then on May 14, 1864, a meteorite blazed across Southern France and fell near the village of Orgueil, scattering stones over a two-square-mile area. Twenty fragments, most the size of a fist, but one as large as a man's head, with a combined weight of 11.523 kilograms were collected. It was a carbonaceous

3 Cm

Figure 27.2. The Orgueil meteorite.

meteorite which disintegrated readily when in contact with water due to the dissolution of water-soluble salts.

The Orgueil meteorite was examined first by S. Cloëz[3] and later by Pierre Berthelot.[4] Berthelot isolated certain organic materials and referred to the "coal-like" substance but was unable to be more specific. The chemical methods of a century ago gave only limited information as to the exact nature and quantities of substances so intractable. Solving the mystery of organic material in meteorites had to wait until recent developments of analytical techniques of organic geochemistry.

The interest in the "bitumin" of meteorites extended from 1834 to 1885; but then followed 68 years when the subject apparently was no longer given any attention by chemists. Finally in 1953, George Mueller[5] at the University of London decided to study the issue. The British Museum with its collection of all 20 of the then known carbonaceous meteorites supplied Mueller with a 20-gram sample of the Cold Bokkeveld, a meteorite that had fallen in the Union of South Africa in 1838.

Mueller attempted first to dissolve the silicates with acid to isolate the organic matter. The process took weeks and was notably more difficult than with terrestrial shale. Mueller found the 1 percent of extractable organic material to be composed principally of carbon, hydrogen, nitrogen, and oxygen, as well as sulfur and chloride. It had a low decomposition temperature. A benzene extract failed to show any optical activity, thus discounting a biological origin. Because it was soluble in alkali, Mueller concluded that the material was composed of complex organic acids with some organic chloride compounds, but he was unable to identify specific chemical structures.

Mueller expressed the view that the organic material must have arisen in an atmosphere of varying illumination and temperature to permit the polymerization of complex molecules. Because it was so difficult to dissolve the silicates, he felt that the organic material probably condensed on dust particles which settled and compacted on a relatively small celestial body. At no time during or after its formation did the organic material experience temperatures greater than 200–350°C. In contrast, the iron and stony iron meteorites had been exposed to temperatures over 1,000°C sometime in their history.

Then in 1960, Brian Mason[6] of the American Museum of Natural History in New York proposed that the primary materials of all meteorites were hydrated silicates covered with a polymerized carbonaceous layer. Wishing to obtain a more specific determination of the organic substances found in the Orgueil meteorite, he sent a sample to chemists at the Esso Petroleum Laboratory in New Jersey for analysis.

Using mass spectroscopy, B. Nagy, W. G. Meinschein, and D. J. Hennessy[7] analyzed the organic portion of the Orgueil meteorite and found an array of hydrocarbons, the largest being a tetracyclic hydrocarbon with a molecular weight of 428. But the most intriguing was an n-paraffin series extending up to C_{26} and peaking at C_{18}. When they compared the spectrum of their meteorite distillate to the mass spectra of butter and recent sediments, they saw a striking similarity. From this they concluded that the meteoritic organic material was from a biogenic source.

That same year Nagy and George Claus[8] of the New York Medical Center examined six carbonaceous meteorites for microscopic detail. The samples were crushed in water or glycerol on glass slides and studied under a light microscope. In two meteorites they found well-defined "organized elements" whose morphology was unlike any known mineral but bore some resemblance to certain species of algae. Since samples from the Orgueil meteorite, which fell in temperate France, and from the Ivuna, which fell 74 years later in an arid tropical region of central Africa, had "organized elements" of similar morphology, Nagy and Claus felt that the probability of the particles being due to terrestrial contamination was unlikely. Consequently, they interpreted these structures as possible microfossils indigenous to the meteorites, and hence, remnants of extraterrestrial life.

Needless to say, this created controversy, and attention was brought to the substances in meteorites. Edward Anders,[9] a chemist at the Enrico Fermi Institute of the University of Chicago, pointed out that museum specimens are routinely labeled with paint, gummed labels, or wax pencil markings, and since the concentrations of hydrocarbons found were of the order of 0.1 to 1.0 micrograms, the possibility of the findings being due to terrestrial contamination was high. But it wasn't the results as much as the conclusion attributing the organic material to extraterrestrial life that was too conjectural.

The controversy at least brought attention to the issue of organic matter in meteorites and expanded the investigation. Other laboratories analyzed extracts from carbonaceous chondrites and found an overwhelming preponderance of aromatic hydrocarbons over the aliphatic. Michael Briggs and G. Mamikunian[10] reported finding 50–90 percent of the material to be aromatic polymer. Martin Studier of the Argonne National Laboratory and Ryoichi Hayatsu and Anders[11] at the University of Chicago analyzed fragments of the Orgueil, Murray, and Cold Bokkeveld meteorites and identified benzene, toluene, naphthalene, anthracene, sulfonic acid esters, and chlorinated hydrocarbons in the organic fraction. They also found the gases H_2, CO, CO_2, NO, N_2, SO_2, CS_2, methane, ethane, and higher homologs. These chemists found no reason to attribute the organic material to a biological origin.

Anders[12] postulated that the meteorites had come from asteroids a few hundred kilometers in diameter that had broken up. For the concentration of the organic material to have been formed by the Miller type of reaction, however, the atmospheres of the asteroids would have had to have been impossibly dense.[13] Rather, it has been shown that the distribution of the aromatic and aliphatic hydrocarbons is in close agreement with the calculations of Dayhoff and her colleagues[14] on the thermodynamic equilibrium for compounds generated in a mixture of carbon, hydrogen, oxygen, and nitrogen at 500°K. That is to say, this organic material was formed and existed in the solar nebula even before the earth came into being.

The study of the meteorites revealed to scientists that components for generating organic substances are not oddities in the universe. It had been known since 1940 that the spectroscopic analysis of light from comets indicated CH and CN radicals. With the development of radio astronomy, astronomers directed their antennas toward the galaxy and beyond and discovered immense clouds of relatively complex molecules in the cosmos. In 1968, A. C. Cheung and others[15] at the radio astronomy laboratory at Berkeley, using a 20-foot radio telescope, detected microwave emissions of clouds of ammonia near the galactic center. The following year the Berkeley group,[16] as well as Lewis Snyder and David Buhl,[17] using the 140-foot radio telescope at Green Bank, West Virginia, detected water. Since then, carbon monoxide,[18] hydrogen cyanide,[19] formaldehyde,[20] cyanoethylene,[21] and other molecular species[22] have been reported. The densities of these molecules in space like that of hydrogen, are small, but the absolute quantities are literally on an astronomical scale.

It was not difficult to accept that the intractable aromatic and aliphatic hydrocarbons of meteorites resulted from a thermodynamic equilibrium of gases of the solar nebula. But that the "organized elements" of Nagy and Claus were remnants of extraterrestrial life was somewhat more difficult. The group at the University of Chicago[23] examined the Orgueil and Ivuna meteorites and failed

Table 27.1. Molecules found in the interstellar medium.

Year	Molecule	Symbol	Wavelength	Telescope	Initial discovery
1937		CH	4300Å	Mt Wilson 100 inch	Dunham
1940	Cyanogen	CN	3875 Å	Mt Wilson 100 inch	Adams }Mt Wilson
1941		CH$^+$	3745-4233 Å	Mt Wilson 100 inch	Adams
1963	Hydroxyl	OH	18, 6.3, 5.0, and 2.2 cm	Lincoln Lab 84 foot	MIT/Lincoln Lab
1968	Ammonia	NH$_3$	1.3 cm	Hat Creek 20 foot	Berkeley
1968	Water	H$_2$O	1.4 cm	Hat Creek 20 foot	Berkeley
1969	Formaldehyde	H$_2$CO	6.2, 2.1 and 1 cm; 2.1 and 2.0 mm	NRAO 140 foot NRAO 36 foot	University of Virginia, NRAO, University of Maryland and University of Chicago
1970	Carbon monoxide	CO	2.6 mm	NRAO 36 foot	Bell Labs
1970	Cyanogen	CN	2.6 mm	NRAO 36 foot	Bell Labs
1970	Hydrogen	H$_2$	1100 Å	UV rocket camera	NRL
1970	Hydrogen cyanide	HCN	3.4 mm	NRAO 36 foot	University of Virginia and NRAO
1970	X-ogen	?	3.4 mm	NRAO 36 foot	NRAO and University of Virginia
1970	Cyano-acetylene	HC$_3$N	3.3 cm	NRAO 140 foot	NRAO
1970	Methyl alcohol	CH$_3$OH	36 and 1 cm; 3 mm	NRAO 140 foot	Harvard University
1970	Formic acid	CHOOH	18 cm	NRAO 140 foot	University of Maryland and Harvard University
1971	Carbon mono-sulphide	CS	2.0 mm	NRAO 36 foot	Bell Labs and Columbia University
1971	Formamide	NH$_2$CHO	6.5 cm	NRAO 140 foot	University of Illinois
1971	Silicon oxide	SiO	2.3 mm	NRAO 36 foot	Bell Labs and Columbia University

Table 27.1. *Continued*

Year	Molecule	Symbol	Wavelength	Telescope	Initial discovery
1971	Carbonyl sulphide	OCS	2.7 mm	NRAO 36 foot	Bell Labs and Columbia University
1971	Acetonitrile	CH_3CN	2.7 mm	NRAO 36 foot	Bell Labs and Columbia University
1971	Isocyanic acid	HNCO	3.4 mm; 1.4 cm	NRAO 36 foot	University of Virginia and NRAO
1971	Hydrogen iso-cyanide	HNC	3.3 mm	NRAO 36 foot	University of Virginia and NRAO
1971	Methyl-acetylene	CH_3C_2H	3.5 mm	NRAO 36 foot	University of Virginia and NRAO
1971	Acetaldehyde	CH_3CHO	28 cm	NRAO 140 foot	Harvard University
1971	Thioformaldehyde	H_2CS	9.5 cm	Parkes 210 foot	CSIRO, Australia

Source: David Buhl, Chemical constituents of interstellar clouds, *Nature 234,* 332–334 (1971). © 1971 Macmillan Journals Limited.

to find the "organized elements," although they did find mineral and sulfur granules bearing a close resemblance. But then Michael Briggs and G. Barrie Kitto[24] of Victoria University in New Zealand reported finding complex organic microstructures in the Mokoia meteorite. They, however, considered the chances of the particles being biogenic as unlikely. Nagy, Claus, and Hennessy[25] countered with another article on the "organized elements" and were answered by a detailed investigation by Fitch and Anders[26] in 1963. The final consensus seems to have been that there were several types of microstructures that fitted as either terrestrial contaminants or structured mineral or structured organic matter, but none indicative of extraterrestrial life.

No sooner had this been concluded than another controversy began to rage over the validity of still other organic compounds detected in carbonaceous chondrites. These were amino acids.

In 1962, E. T. Degens and M. Bajor[27] at the California Institute of Technology found sugars and amino acids in an 80 percent ethanol extract of the Murray and Bruderheim meteorites. A detailed article the following year by I. R. Kaplan, Degens, and J. H. Reuter[28] gave the results of an analysis of eight carbonaceous chondrites and five noncarbonaceous meteorites. Amino acids and sugars were found in all of them. The sugars were mannose and glucose in concentrations of 5–26 micrograms per gram; and 17 amino acids were found in amounts of

30–500 micrograms per gram for the carbonaceous meteorites, 8–50 micrograms for the noncarbonaceous. Since the organic material lacked optical activity, as well as pigments, fatty acids, and presumably nucleic acids, the authors were of the opinion that the sugars and amino acids were of a chemical rather than a biological origin.

The quantities of the amino acids were of the order of 10^{-6} to 10^{-8} moles per gram. Biochemists who run paper chromatograms to detect amino acids with ninhydrin spray are aware that the method is sensitive enough to give a positive test for amino acids from fingerprints if proper care is not exercised. Paul Hamilton[29] at the Dupont Institute in Wilmington, Delaware, made a systematic study of the amount of amino acids that can result from handling. A single print with a dry thumb from hands washed 2 hours previously was pressed on the inside wall of a dry beaker. The print was washed into the beaker, the water evaporated, and the residue chromatographed on an ion-exchange column in the routine procedure. Seventeen amino acids of the order of 5×10^{-9} to 2×10^{-10} moles were detected. Serine was in the largest amount, followed by glycine. Juan Oró and H. B. Skewes[30] also analyzed amino acids from fingerprints and compared the results to those found in meteorites. The amounts were of the same order of magnitude and the ratio of the amino acid pairs Ser-Thr, Gly-Ala, and Ser-Ala were in close agreement.

Since the stones had been handled and kept in museums, there was serious doubt as to the validity of amino acids in meteorites. Nonetheless, the issue was not closed. The researchers who reported the findings were aware of the difficulties involved with low levels of substances for analyses and had confidence in their procedures. Fresh evidence was needed in which there could be no question of terrestrial contamination to settle the issue one way or the other.

The opportunity came from the sky at 11:00 A.M. on September 28, 1969, when a carbonaceous chondrite fell near Murchison, Victoria, Australia. The parent object broke during the flight and scattered many fragments over a five-square-mile area. Specimens were collected soon after the fall and analyzed jointly by NASA scientists at Ames Research Center in California, Isaac Kaplan at Berkeley, and Carlton Moore of the Arizona State University.[31]

The Murchison meteorite contained 2 percent by weight of carbon and 0.16 percent nitrogen. When a 10-gram sample was hydrolyzed with hydrochloric acid and analyzed by chromatography, it gave peaks for glycine, alanine, valine, proline, and glutamic acid. In addition to these five amino acids common to proteins, there were 2-methylalanine and sarcosine, not normally found in biological systems. In this analysis there was sufficient material to measure optical isomers, and the D and L forms were found to be in nearly equal proportions. Terrestrial organisms use almost exclusively the L isomer. The results clearly established that amino acids in meteorites were not present as a result of contamination and that they were apparently of nonbiogenic origin. A second

Figure 27.3. The Murchison meteorite.

report[32] identified 11 amino acids in all in the meteorite. The Murchison meteorite contained a total of 2×10^{-7} moles of amino acids per gram—a content higher than that contained in many desert sands.

The absence of optical activity with extremely old amino acids does not rule out the fact that they may have been of only one isomeric form at one time. The stereoisomerism of amino acids is not absolutely stable. Amino acid isomers will racemize eventually at a rate dependent on the temperature—a feature that has been developed as a means of dating ancient speciments.[33] At 0°C the isomers of isoluecine and alanine have a half-life for converting to a 50–50 mixture of 4.4 and 1.1 million years, respectively, whereas at 25°C their half-lives are 35,000 and 11,000 years.

Meteorites are the oldest rocks known. They have been dated at 4.6 billion years and are presumed to be as old as the solar system itself. This leaves the question of the origin of the meteoritic amino acids. Were they generated by the Miller type of reaction in the solar nebula or could they have come about in some other manner? It is possible that they came from an unforeseen source. When rocks brought back from the moon by the Apollo missions were analyzed for amino acids, to the surprise of many scientists, the initial results were negative. But these analyses were carried out on simple water extractions of the samples. When it was found that there were no amino acids in the water extracts, the extracted material was subjected to hydrolysis in the routine manner for proteins, then analyzed. Six amino acids—glycine, alanine, aspartic acid, glutamic acid, serine, and threonine—were detected. The amounts were small—7 to 45 nanograms per gram.[34]

Why then should amino acids be present only after hydrolysis? Apparently on the moon in the absence of water, the amino acids do not exist, but their precursors do. The hydrolysis converted these precursors to free amino acids. Nothing is known of the chemical nature of these precursors, but it is conjectured that they may be cyanides—cyanides that resulted from the ceaseless bombardment of the lunar surface by carbon and nitrogen nuclei of the solar wind.

28
Gaia

Life began nearly 4 billion years ago in what seems like a strange and alien setting. Life began, and with it the long course of biological evolution. Across the vast, incredible expanse of time countless generations formed, reproduced, and died in mindless existence, changing the chemical nature of the earth, changing themselves progressively on an evolutionary scale, rung after rung, from microbe to man. The most astonishing result of the whole process is that we know about it.

We are, in effect, chemical substances that came together into a self-sustaining, replicating system and evolved through the eons into the light of consciousness. The consciousness progressed to self-awareness, a state in which we can, as products of the process, look back into time and reconstruct the conditions of our own beginning.

That this should happen with us is not coincidental. Human self-awareness created an ambition to know the world and change it to satisfy our desires. It is a feedback process, like the autocatalysis of the emerging biological cell that set the system apart from the surrounding milieu. Self-awareness severed man's total dependence on the vagaries of the environment and made mankind a particular stage in evolution.

Because of our ability to see ourselves in the physical surroundings, we have made an impact on earth like no other living kind. We can look at the various forms of life, past and present, and see their place in the tight knit of nature's fabric. But technological man has broken the bonds that hold other creatures to ecological confinement. Instead of being pressed into the bionomic mold by the Darwinian precept of natural selection, we have achieved the wherewithal to shape nature to our casting. And as man advances in technology, his presence seems to take on a significance that extends beyond the terrestrial to the cosmic, for he alone of the living has managed to free himself of earth's embrace.

Our technological achievements have given us the means to expand our niche from Earth into the solar system. Mars lies 60 million miles away. It is 600 million miles to the moons of Jupiter, 1 billion miles to Saturn. Uranus, Neptune, and Pluto remain. But there ends our system. Beyond stretches an expanse of space so immense that it takes years for light to traverse it from the nearest stars. The microbe that struggled through 3.5 billion years of evolution to become man looks beyond to the 100 billion stars in the galaxy and discovers that he has broken out of one droplet universe only to find himself in another.

Yet, we know the microbe lives in blind isolation in a microcosm that is only part of a vast dimension. We know that animals of a simpler nature than man live on the periphery of a physical reality with depth and complexity far beyond their ability to comprehend. Where then lies man? Are we too on the edge of a reality that reaches into domains of existence we can suspect but not understand or is man in an evolution that is itself a quality of the material world that extends without limit?

The tide that has carried all existence along on its inexorable flow continues. It does not terminate with us. We are merely participants on a multiphasic stage, centered in a narrow slit of time and space, while the cascade of events that began so many eons ago tumbles on in a relentless flow. We are a part of a process that began in some Archean catchbasin, and from that humble beginning, we climbed tier after tier on what seems to be a step pyramid. But now we seem to be reaching the apex—a point where evolution's trend proceeds no further. On analysis of life's direction, we discover that it isn't man that has arrived at a critical juncture of a long journey—it is biological evolution.

Mutation created the variety, and natural selection made the fit for organisms to expand into every ecological niche accessible to them. But the biological field lies strewn with species that remain on levels of past existence. Evolution followed a series of steps to greater size and complexity, while the various levels remained occupied by arrested forms of life. Only segments of biology evolved and made the step-wise climb in succession to man.

What impetus, therefore, caused the descendants of the procaryotes to leave their microbial world in which they had flourished for so many eons to venture into a whole new dimension?

Size and complexity are not achieved in evolution solely by the selective mechanism of Darwinsim. Within a level of evolutionary development the tenets of Darwinian evolution remain valid. But expansion into a hierarchy of stages requires organizing more matter in highly ordered compositions. There seems to be, therefore, some underlying principle which pushes the column of biological systems to ever greater heights. To find it we must reexamine the steps.

The largest step in evolution is one of the least apparent to the casual observer.

Over a thousand million years ago during the Precambrian, when the continents stood stark and barren, when life was still confined to microorganisms floating in the oceans, the eucaryotes appeared—larger and considerably more sophisticated in internal structure than their procaryotic ancestors. The eucaryotes were still single-celled microbes, but they had evolved an enzymic pathway that allowed them to use molecular oxygen to oxidize glucose completely to carbon dioxide and water. What the eucaryotes achieved by respiration was a release of available energy from glucose that was 18 times greater than their predecessors were able to obtain.

When the eucaryotes evolved respiration, they ushered in an era of multicellularity that thrust plants and animals into a whole new dimension. The principal steps in evolution since then have been attributable to modification of this system for greater metabolic efficiency. Fish evolved with a more efficient circulatory system, using hemoglobin for extracting oxygen from water to carry to all cells of the body. Reptiles were air breathers that could capitalize on the greater concentration of oxygen in the atmosphere. Mammals appeared with temperature control to keep the blood warm and thus more efficient in supplying the brain and muscles with oxygen. Man branched from the primates by diverting motivation energy through sublimation to creative thought, the step that carried him across the watershed to tap the vast reservoirs of energy found in inanimate nature.

There is one single factor that can be identified with each major step—energy.

Whether for the growth of a biological system or the construction of a skyscraper, arranging matter in a more ordered composition counter to the universal tide takes an energy concentration. The greater the extent of material organization, the greater the flood of energy needed from some reservior to drive the construction. Each major advance in the evolutionary column has been accompanied by the development of a means to draw upon energy that was unavailable to the old order.

Energy is the springboard that thrusts life forms into greater size and complexity, into larger physical reality. And as we arrange the sequence of evolution's advance, we discover an unsettling implication.

Each step is an evolutionary curve; all steps together outline an accelerating advance for all biological evolution. Over half of the time was used to advance from procaryote to eucaryote. It took half again the time to reach the level of fish. And as the succeeding steps followed, the succession time shortened. It is the curve of an accelerating object building momentum, like a ball dropped from a height. The driving force goes unchecked; momentum sets the pace.

Each major development in evolution appears to take less and less time to occur. And each development begins slowly but, fed by its own momentum, begins to accelerate until it races to its developed state. When it reaches a final level—a higher stage in evolution—the offspring of the new life form begin to repeat the cycle, evolving some feature that ultimately leads to another succeeding step.

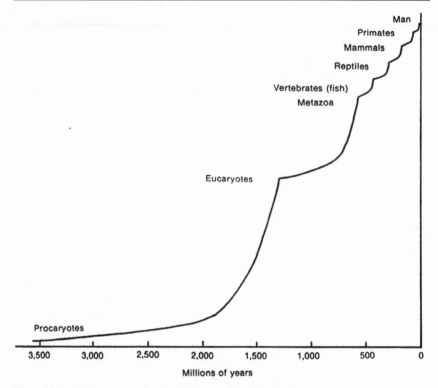

Figure 28.1. Diagram tracing the principal stages of evolution leading to man.

Segments of biology equilibrate and stop evolving, but the overall advance of the column does not reach equilibrium. To the contrary, it continues to accelerate stage after stage to such a rate that it suggests that the interval for man's preeminence will be ominously short. We apparently have reached a critical point in biological evolution. Either the trend of evolution is no longer valid or a radical change in the evolutionary process is imminent. In any event, we are in the middle of something momentous that is taking place.

The machinery of biological systems runs on spontaneous reactions—chemical reactions that are out of equilibrium with the surroundings and represent an energy potential as a driving force. By isolating and coordinating select chemical processes within an insulating membrane, a cell was able to create its own energy potential and direct its force to carry out cellular activities.

The process of selectively isolating the cell from the environment by a membrane was essential for the potential. It also made the primeval cell an independent entity; and when the cell became autocatalytic, it broke from its heavy dependence on the chemical environment for the control of its own processes and evolved away from the prebiotic milieu toward a separate development.

These two steps—the formation of an energy potential as a driving force and the use of cyclical processes to create independent stages—have been major procedures in evolution. They were essential for the formation of biological cells; they are directly related to man's preeminence. Self-awareness has been the cyclical reaction that has allowed us to see ourselves independent of the natural scene. And the use of external energy has been responsible for man's ascent away from animality.

Our progress in the control of energy traces the path of our technological society. The first step was a long one. It began with Prometheus, but remained a slow development for a long time. For half a million years fire was man's only external energy of importance. During that long interval he improved on weapons and tools, but they remained as extensions of his metabolic energy. Only around 10,000 years ago did agriculture begin to replace the hunting society. Communities took root and grew; beasts of burden supplemented human labor. As engineering principles were discovered, the flow of the wind and water were drawn upon for work, but metabolic energy remained as the principal source of controlled energy. The great potential of fire to perform work was not harnessed until the steam engine was invented less than 200 years ago. As the wheels of industry began to turn, the energy sealed in the vast reserves of fossil fuels was tapped. Then in 1945, the energy of the atom was released under controlled conditions in an atomic reactor at the Argonne Laboratories in Chicago.

As social and technical advancements prepared the setting, energy and our control of it became increasingly more important. The pace quickened with each step, until now society consumes energy at an astronomical rate. The development traces a typical curve of evolutionary change where the advance begins slowly, then accelerates, until toward the end it races to another organizational level.

But what organizational level could possibly succeed us?

Individual species and their environments cannot be regarded in isolation. Each ecological system is a part of a series of ecological systems that enlarges by stages from the smallest to the largest through interconnected feedback loops until they include the entire biosphere. Nor is the environment separate from the living creatures that inhabit it. The soil, the sea, the air are a part of the biosphere that they nourish and are affected by the equilibrium between the gases of the atmosphere and the oceans, and there are biological gas exchange processes between the atmosphere and all living things. Plants take up carbon dioxide and release oxygen; animals use oxygen and expel carbon dioxide. In addition, there are a number of gases in lesser amounts, such as methane, nitrous oxide, carbon monoxide, and ammonia, that are released into the atmosphere by biosystems, particularly by microorganisms. The atmosphere consists, therefore, of approximately a four to one ratio of nitrogen to oxygen, with a large number of minor constituents in concentrations that stay surprisingly constant over long periods of geological time. The point of interest is that they are all far out of equilibrium.

The earth's atmosphere is an oddity in a universe that is 75 percent hydrogen.

An atmosphere consisting of 21 percent oxygen represents a large energy potential with the earth's interior. This potential is the result of and is sustained by biological processes. Without plants releasing oxygen into the atmosphere from photosynthesis, it has been estimated that essentially all of the earth's free oxygen would be lost within 2,000 years to the oxidation of rocks and gases from volcanic activity.[1]

Not only are the two major gases nitrogen and oxygen out of equilibrium, but the minor atmospheric components also are far more abundant than they should be according to equilibrium chemistry. Moreover, the atmosphere's composition remains relatively constant, despite the assaults by nature and industrial man. There must be some regulating mechanisms that maintain this constancy. And to sustain the disequilibrium, the mechanisms have to be continually charged by an input of energy.

We identify life with biological systems. It is the unique quality of a cell of biological molecules encapsulated in a semipermeable lipid membrane. This qualilty was passed on to the complex plants and animals when cells crossed over to multicellularity. The quality, however, has a wider implication than the narrow definition restricted to discrete organisms. Combinations of organisms, such as societies of microorganisms, insects, and man take on a vitality of their own. They draw from the energy of their members and funnel it into activities of the whole unit. It is in turn a symbiotic relationship where the individuals derive benefits they could not attain alone. The dependence is life-sustaining. Neither ants, nor bees, nor man can live for long as solitary creatures. When the society dies, the members die like cells of a deceased body.

The quality we call life is created in systems that are self-sustaining and able to maintain an energy potential to drive their activites. The potential is a disequilibrium across a boundary between the unit's interior and the surrounding environment. For a biological cell the boundary is the lipid membrane; for a society it is set by the closed membership; for the entire biosphere it is the oxygenated atmosphere and the sum of the potentials of the biosphere that extends the imbalance.

When we look at the earth we discover that the entire system has these same characteristics of a viable entity.

James Lovelock of England and Lynn Margulis at Boston University have advanced the Gaia Hypothesis,* in which the earth's atmosphere is seen as the circulatory system of the biosphere.[2] Certain aspects of the atmosphere—temperature, composition, oxidation–reduction state, acidity—form a homeostatic system. Since these properties are themselves products of evolution, the atmosphere has the appearance of being a contrivance formed collectively by living systems to carry out necessary control functions. Living matter, the air, the oceans, and the land surface appear as parts of a gigantic system which is able

---

* Gaia—the ancient Greek goddess of the earth.

to control its temperature, the composition of the sea and air, and the acidity of the soil for survival of the biosphere. In other words, the entire system seems to behave like a living organism.

We are a process within a process. Since Prometheus we have been on an accelerating curve, unleashing reserves of energy to build our dominance over nature. But we are not an isolated phenomenon. We are the culmination of a biological evolution that has transformed the earth into a dynamic process of awesome proportions. And now we have come a full cycle. Our position is strikingly similar to that of the eucaryotic organisms when they emerged over a thousand million years ago and broke through to a whole new dimension.

For 2 billion years cyanobacteria had formed the base for absorbing solar energy. The biosphere assimilated this energy internally by lateral growth and by pyramidal growth of food chains, while the algae liberated free oxygen into the environment. Mutational change, like ticks in a clock to higher organization of DNA, was sparked by the energetic impingement of cosmic rays; natural selection worked as the ratchet against reverting to a lower level.

But the mechanism marked time. Life remained on the level of procaryotes until the entire terrestrial environment could be brought forward. It took eons, but respiration was not possible until the free oxygen level reached 1 percent of today's value. It was at this point that the energy potential was high enough to be tapped by emerging eucaryotic cells for greater structural organization. So great was the energy potential that the door opened to a whole new dimension in size, and the eucaryotes passed through on their march to higher levels.

The crossover from the unicellular stage to the tissue level of development was made with sponges and coelenterates. It took another 100 million years to advance from the tissue level to animals with organs. But the result was explosive. Undaunted and unopposed, biology rushed into the dimensional void with the rapid evolution of all but one animal phyla (basic anatomical types). Only the vertebrates were to evolve later from the chordates. Arthropods, mollusks, worms, and other invertebrates populated the seas, and biological evolution commenced on the road to repeat the cycle.

For 600 million years evolution has been drawing from the earth's energy potential to expand the biosphere throughout the new dimension.* Food chains evolved to accommodate the bulging energy reserve. The biological structuring of animal life became extremely complex and refined. Every conceivable ecological niche from the frigid waters of the polar regions to the oceans' abyssal depths has been occupied. Man in his expansion extended his dominance into the domains of all other living creatures. But more significantly, the children of Prometheus have opened Pandora's box and found energy hoards of unimaginable potential. They have released the vast reserves of energy in fossil fuels, and now the awesome energy of the atom.

* The potential is maintained by photosynthesis, especially by the phytoplankton in the oceans.

Man's biological evolution reached a plateau around 100,000 years ago and has remained little changed since. Our development as a species has matured; it is our cultural structure that continues to evolve. Our biology has equilibrated but our technological society has not. It continues to race with increasing momentum on an accelerating curve, fueled by the impetus of energy that goes unopposed. We use the energy to build, refine, and rebuild the order of matter around us. We are in a stage that is equivalent to the period that preceded the eucaryotic revolution. The procaryotes had expanded in number and evolved to the limit of their sophistication. Their use of energy, however, did not lead to equilibrium. Instead, it accelerated the energy potential that was building and led to their loss of preeminence.

Our technology consumes energy at an enormous rate and in the process, devises means to uncover even greater sources. But there is a limit to how much energy can be absorbed in an orderly manner within a confined space. As happened with the eucaryotes, our potential will catapult us into a greater dimension.

We look out through the thin gaseous envelope that sheaths the earth and see a cosmos sprinkled with billions upon billions of other worlds. They lie before us like distant shores, like continents unexplored. Space is vacant, the lands are barren, inviting, like a repetition of that event at the dawn of the Cambrian so many hundred million years ago when our ancestors broke through to another dimension and came out of the sea to colonize the land. The prospects fire our imagination and grip our destiny. The vastness of space holds us in awe and slowly pulls us toward it like an irresistible magnet.

Our expansion has repeatedly had to overcome the combined obstacles of time and distance. While our time sense has remained relatively unchanging within a narrow range, we have breached the barriers by shrinking space with faster movement. But the cosmos is another dimension. The distances to the stars are so enormous by our standards that it is no longer possible within our physical laws to make the crossing in our time. We have reached the end of the tether for distending our segment of the material web. Bound to limits of a size-space position, we cannot make movement great enough to connect our system with another stellar position. To do so we have to enlarge our size-space dominance to cosmic proportions.

We are at the interface between two great dimensions—the microcosm and the cosmos. One is structured on the force of electrostatic attraction with the molecular and cellular architectures as outgrowths of it. The other is patterned on the force of gravity between large masses. Size is the distinction between the two dimensions and runs inversely with the strength of the binding force. A microbe, surrounded by and closer to the strong actions of the molecular world, is below the touch of gravity and unaware of its existence. But we are at the interface. We see those things smaller than us responding to chemical affinities, and our mass feels the pull of gravity and relates to us the webbing that holds the cosmos.

Our biological bias blinds us to our cosmic role. We regard life and consciousness as the ultimate forms of material realization and believe that man and his Darwinian successors will lead a conquest of space. We entertain thoughts of communicating with other systems of our kind in the galaxy, to bring them and ourselves out of dire isolation into a common light of conscious existence. But the stars are the loci to the webbing of the cosmic dimension. Its beaches are planets; its continents are planetary systems; its worlds are island universes of billions of stars scattered in a vacuous sea of subatomic nuclei and particles. To enter and become a part of this arena we have to have the size, the complex composition, and the command of a corresponding energy to control matter and space on that scale.

The procaryotes created an energy potential they could not handle. They were succeeded by not a more evolved form of procaryote but an entirely new creation that arose from a consortium of diverse types. In the variety among the countless billions of procaryotic microbes were organisms with enzymes to use molecular oxygen for respiration. Others had evolved with enzymes to protect them from destruction by oxidation. When the potential of atmospheric oxygen reached a critical level during the Precambrian period, the two types of organisms formed a symbiotic union and became one. The procaryotes then retreated to haunts of the environment to live out their existence as living fossils of a bygone age. Today, the human body alone harbors vast numbers of bacteria that have adapted themselves to live within an evolved product of their own creation.

And now the cycle is approaching a full turn. The procaryotes lost their preeminence not solely because they poisoned their environment but because they were on the edge of a larger dimension and created an energy potential that could propel a greater form of life into it. We, too, stand poised on a springboard, surveying the direction of our future.

We have the illusion that our actions are our own, that it is our hand that holds the latchkey. But the universe does not share our biological bias. Life and consciousness are properties of biocreatures. In a cosmic evolution, biology is a phase that lies between chemical evolution and an evolution that is to follow. Our biological characteristics are qualities that evolved for and fit the conditions of our dimension. The dimension of the cosmos is for systems that can transcend the time–space expanses of galactic space.

As we peer out ever farther through the crack in our shell of egocentricism into time and space, we see a universe swirling in dynamic evolution. And we realize that perhaps, just perhaps, we have been seeking the answer to the wrong question. No longer is it a question of how the universe fits into the being of man. How does man fit into the being of the universe?

Mankind alone is not reaching into space; it is the evolving earth and its biosphere that is beginning to extend itself and its projections into the solar system. Man's technology represents the end of biological evolution's extent and the initiation of a mechanical evolution. Our devices are at the stage that chemical evolution was on primordial earth just before it created the first cells and ushered in the biological evolution. Machines are on the path toward self-sufficiency, like the biological cells. We may even build into them our consciousness and make them sufficiently independent that we inadvertently launch them on their own evolution.

Our creations can lead to a network the size of the solar system. Like a growing embryo, where cells differentiate to serve a particular function, our technological direction can grow into the solar system with humans and machines in union. Eventually man's role may be phased out and he will remain in an arrested state of development like the procaryotic microbe in his droplet universe, while his creation goes on to greater dimensions.

As we play out our lives as individuals and as a species, our role in the cosmic process seems predestined. There is a cosmic significance to man's direction. In an exploding universe racing toward chaos, we try to stem the tide by compressing time and space back again into order. The cataclysmic burst of energy that created the material universe has become a cooling broth seasoned with organizing potential. Like the legendary phoenix that arose renewed from its own ashes, the universe has within it the seeds of its own resurrection.

We may never know what we create, may never know the qualities that lie past consciousness and self-awareness. The advance into a larger and more evolved form will carry our creation into qualities beyond the pale of human comprehension—beyond our segment of the universe to a new dimension and a new reality.

Notes

Chapter 1. Building Blocks

1 H. Urey, *The Planets* (New Haven: Yale University Press, 1952).
2 H. C. Urey, On the early chemical history of the earth and the origins of life, Proc. Nat. Acad. Sci. *38*, 351–363 (1952).
3 A. I. Oparin, *The Origin of Life*, trans. S. Morgulis (New York: Macmillan, 1938).
4 S. L. Miller, A production of amino acids under possible primitive earth conditions, Science *117*, 528–529 (1953).
5 S. L. Miller, The first laboratory synthesis of organic compounds under primitive earth conditions, in *The Heritage of Copernicus: Theories Pleasing to the Mind*, J. Neyman, ed. (Cambridge, Mass.: The MIT Press, 1974), pp. 228–242.

Chapter 2. Early Earth

1 B. G. Marsden and A. G. W. Cameron, *The Earth-Moon System* (New York: Plenum Press, 1966), p. 73.
2 William W. Rubey, Geologic history of seawater, Bull. Geol. Soc. Am. *62*, 1111–1147 (1951).
3 S. Moorbath, R. K. O'Nions, and R. J. Pankhurst, Early Archean age for the Isua Iron Formation, West Greenland, Nature *245*, 136–139 (1973).
4 L. Paul Knauth and Samuel Epstein, Hydrogen and oxygen isotope ratios in nodular and bedded cherts, Geochim. Cosmochim. Acta *40*, 1095–1108 (1976).

Chapter 3. Life before the Precambrian

1 M. F. Glaessner, Pre-Cambrian animals, Sci. Amer. *204* (*3*), 72–76 (1961); R. Goldring and C. N. Curnow, The stratigraphy and facies of the Late Precambrian at Ediacara, South Australia, J. Geol. Soc. Aust. *14*, 195–214 (1967); M. Wade, Preservation of soft-bodied animals in Precambrian sandstones at Ediacara, South Australia, Lethaia *1*, 238–267 (1968).
2 Adolph Knopf, The boulder batholiths of Montana, Amer. J. Sci. *255*, 81–103 (1967).
3 M. R. Walter, *Stromatolites and Biostratigraphy of the Australian Precambrian and Cambrian*, (London: The Palaeontological Association, 1972).
4 Preston Cloud and Aharon Gibor, The oxygen cycle, Sci. Amer. *223* (*3*), 110–123 (1970).
5 P. Cloud, Evolution of ecosystems, Amer. Scientist *62*, 54–56 (1974).
6 S. S. Gildrich, Ages of Precambrian banded iron-formation, Econ. Geol. *68*, 1126–1134 (1973).
7 S. Moorbath, R. K. O'Nions, and R. J. Pankhurst, Early Archean age for the Isua Iron Formation, West Greenland, Nature *245*, 138–139 (1973).
8 S. A. Tyler and E. S. Barghoorn, Occurrences of structurally preserved plants in Precambrian rocks of the Canadian Shield, Science *119*, 606–6–8 (1954).

9 E. S. Barghoorn and S. A. Tyler, Microorganisms from the Gunflint chert, Science *147*, 563–577 (1965).

10 J. W. Schopf, Microflora of the Bitter Spring Formation, Late Pre-Cambrian, Central Australia, J. Paleontol. *42*, 650–688 (1968).

11 E. S. Barghoorn and J. W. Schopf, Alga-like fossils from the Early Precambrian of South Africa, Science *156*, 508–512 (1967).

12 Hans D. Pflüg, Structured organic remains from the Fig Tree Series (Precambrian of the Barberton Mountain Land, South Africa), Rev. Palaebot. Palynol. *5*, 5–29 (1967).

13 A. Engel, B. Nagy, L. A. Nagy, E. G. Engel, G. O. W. Kremp, and C. M. Drew, Algal-like forms in Onverwacht Series, South Africa: oldest recognized life-like forms on earth, Science *161*, 1005–1008 (1968).

14 J. Brooks and M. D. Muir, Chemistry and morphology of the micro-organisms in the Early Precambrian rocks of the Onverwacht Group, I.U.P.A.C. International Symposium on Chemistry in Evolution and Systematics, held at Strasbourg, France, July 3–8, 1972.

15 J. Oró and D. W. Noones, Aliphatic hydrocarbons in Precambrian rocks, Nature *213*, 1082–1083 (1967).

16 J. Han and M. Calvin, Occurrence of fatty acids and aliphatic hydrocarbons in a 3.4 billion-year-old sediment, Nature *224*, 576–577 (1969).

17 B. Nagy, Porosity and permeability of the Early Precambrian Onverwacht chert origin of the hydrocarbon content, Geochim. Cosmochim. Acta *34*, 525–527 (1970).

18 D. Z. Oehler, *Carbon Isotopic and Electron Microscope Studies of Organic Remains in Precambrian Rocks* (Ph.D. thesis, Univ. of Calif., Los Angeles, 1973).

19 D. Z. Oehler, J. W. Schopf, and K. A. Kvenvolden, Carbon isotopic studies of organic matter in Precambrian rocks, Science *175*, 1246–1248 (1972).

20 J. W. Schopf, Biogenicity and significance of the oldest known stromatolites, J. Paleontol. *45*, 477–485 (1971).

21 J. S. R. Dunlop, M. D. Muir, V. A. Milne, and D. I. Groves, A new microfossil assemblage from the Archaean of Western Australia, Nature *274*, 676–678 (1978).

22 A. S. Lopuchin, Structures of biogenic origin from Early Precambrian rocks of Euro-Asia, Origins of Life *6*, 45–47 (1975).

23 J. William Schopf, Are the oldest "fossils," fossils? Origins of Life *7*, 19–36 (1976); Preston Cloud and Karen Morrison, On microbial contaminants, micropseudofossils, and the oldest records of life, Precambrian Res. *9*, 81–91 (1979).

24 P. E. Cloud, G. R. Licari, L. A. Wright, and B. W. Trowel, Proterozoic eucaryotes from eastern California, Proc. Nat. Acad. Sci. *62*, 623–630 (1969).

25 J. William Schopf and Dorothy Z. Oehler, How old are the eucaryotes? Science *193*, 47–49 (1976).

26 Gerald R. Licari, Biogeology of the late pre-Phanerozoic Beck Spring Dolomite of eastern California, J. Paleontol *52*, 767–792 (1978).

27 G. R. Licari and P. Cloud, Prokaryotic algea associated with Australian Proterozoic stromatolites, Proc. Nat. Acad. Sci. *62*, 56–62 (1972).

28 H. Tappan, Possible eucaryotic algea (Bauglophycidae) among early Proterozoic microfossils, Geol. Soc. Am. Bull. *87*, 633–639 (1976).

29 A. H. Knoll and E. S. Barghoorn, Precambrian eukaryotic organisms: a reassessment of the evidence, Science *190*, 52–54 (1975).

30 B. Bloeser, J.W. Schopf, R. J. Horodyski, and W. J. Breed, Chitinozoans from the Late Precambrian Chuar Group of the Grand Canyon, Science *195*, 676–679 (1977).

Chapter 4. The Age of Procaryotes

1 D. Z. Oehler, J. W. Schopf, and K. A. Kvenvolden, Carbon isotopic studies of organic matter in Precambrian rocks, Science *175*, 1246–1248 (1972).

2 Josephine E. Tilden, *The Algae and Their Life Relations* (Minneapolis: University of Minnesota Press, 1935).

Chapter 5. The Advance of the Eucaryotes

1 L. Paul Knauth and Samuel Epstein, Hydrogen and oxygen isotope ratios in nodular and bedded cherts, Geochim. Cosmochim. Acta *40*, 1095–1108 (1976).
2 B. G. Marsden and A. G. W. Cameron, *The Earth-Moon System* (New York: Plenum Press, 1966), p. 73.
3 Dorothy Hinslow Patent, *Microscopic Animals and Plants* (New York: Holiday House, 1974), p. 61.
4 H. V. Wilson, On some phenomena of coalescence and regenerative sponges, J. Exptl. Zool. *5*, 245–258 (1907).
5 J. Holtfreter, Gewebeaffinität ein Mittel der embryonaten Formbildung, Arch. exptl. Zellforsch. Gewebezücht *23*, 169–209 (1939).
6 A. Moscona and H. Moscona, The dissociation and aggregation of cells from organ rudiments of the early chick embryo, J. Anat. *86*, 287–301 (1952).

Chapter 6. Life's Cellular Nature

1 J. Cairns, from Braum: *Bacterial Genetics*, 2d ed. (Philadelphia: W. B. Saunders Co., 1965.
2 B. Byers, Structure and formation of ribosome crystal in hyperthermic chick embryo cells, J. Mol. Biol. *26*, 155–167 (1967).
3 Ernest Borek, *The Sculpture of Life* (New York: Columbia University Press, 1973), pp. 10–12.
4 J. M. Whatley, The fine structure of Prochloron, New Physiologists *79*, 309–311 (1977).

Chapter 7. Molecular Architecture

1 J. B. Sumner, Isolation and crystallization of the enzyme urease, J. Biol. Chem. *69*, 435–440 (1926); J. B. Sumner, Recrystallization of urease, J. Biol. Chem. *70*, 97–98 (1926).

Chapter 8. The Molecular Basis of Life

1 F. Miescher, Über die chemische Zusammensetzung der Eierzellen, Hoppe-Seyler Med. Chem. Untersuch 44a (1871).
2 O. T. Avery, C. M. MacLeod, and M. McCarty, Studies on the chemical nature of the substance inducing transformation of pneumococcal types, J. Exptl. Med. *79*, 137–157 (1944).
3 G. W. Beadle and E. L. Tatum, Genetic control of biochemical reactions in *Neurospora*, Proc. Nat. Acad. Sci. *27*, 499–506 (1941).
4 W. M. Stanley, Isolation of a crystalline protein possessing the properties of Tobacco Mosaic virus, Science *81*, 644–645 (1935).
5 E. Schrödinger, *What Is Life?* (New York: Macmillan, 1945).
6 E.Chargaff, S. Zamenhof, G. Bravermian, and L. Kerin, Bacterial deoxypentose nucleic acids of unusual composition, J. Am. Chem. Soc. *72*, 3825 (1950).
7 L. Pauling and R. B. Corey, Structure of the nucleic acids, Nature *171*, 346 (1953); L. Pauling and R. B. Corey, A proposed structure for the nucleic acids, Proc. Nat. Acad. Sci. *39*, 84–97 (1953).
8 J. D. Watson. *Molecular Biology of the Gene*, 2d ed. (New York: Benjamin, 1970); J. D. Watson and F. H. C. Crick, A structure for deoxyribose nucleic acid, Nature *171*, 737–738 (1953); J. D. Watson and F. H. C. Crick, Genetic implications of the structure of deoxyribonucleic acid, Nature *171*, 964–967 (1953).
9 M. Goulian, A. Kornberg, and R. L. Sinsheimer, Enzymatic synthesis of DNA. XXIV. Synthesis of infectious phase Ø X-174, Proc. Nat. Acad. Sci. *58*, 2321–2328 (1967).

Chapter 9. From Blueprint to Organism

1 T. Caspersson, Studien über den Eiweissumsatz der Zelle, Naturwiss. *29*, 33–48 (1941).
2 J. Bracht, La localization des acides pentosenucleiques dans les tissus animaux et les oeufs d'Amphibiens en voie de developpement, Arch. Biol. (Liege) *53*, 207–257 (1942).
3 H. Borzook, C. L. Deasy, A. J. Hagen-Smit, G. Keighley, and P. H. Lowry, Metabolism of C¹⁴-labeled glycine, L-histidine, L-leucine and L-lysine, J. Biol. Chem. *187*, 839–848 (1950).
4 R. W. Holley, J. Aggar, G. A. Everett, J. T. Madison, M. Marquisse, S. H. Merrill, J. R. Penswick, and A. Zamur, Structure of a ribonucleic acid, Science *147*, 1462–1465 (1965).
5 J. T. Madison, G. A. Everett, and H. Kung, Nucleotide sequence of a yeast tyrosine transfer RNA, Science *153*, 531–534 (1966).
6 J. D. Watson, Involvement of RNA in the synthesis of proteins, Science *140*, 17–26 (1963).
7 M. W. Nirenberg and J. H. Matthael, The dependence of cell-free protein synthesis in E. coli upon naturally occurring or synthetic polynucleotides, Proc. Nat. Acad. Sci. *47*, 1588–1602 (1961).

Chapter 10. A Thread Unbroken

1 E. Zuckerkandl and L. Pauling, Evolutionary divergence and convergence in proteins, in *Evolving Genes and Proteins*, V. Bryson and H. J. Vogel, eds. (New York: Academic Press, 1965), pp. 97–166.
2 Motoo Kimura, The rate of molecular evolution considered from the standpoint of populations genetics, Proc. Nat. Acad. Sci *63*, 1181–1188 (1969).
3 Vernon M. Ingram, Gene evolution and hemoglobins, Nature *189*, 704–708 (1961).
4 *Atlas of Protein Sequence and Structure*, vol. 5, M. O. Dayhoff, ed. (Washington, D.C.: National Biochemical Research Foundation, 1972).
5 W. M. Fitch and E. Margoliash, Construction of phylogenetic trees, Science *155*, 279–284 (1964).
6 Motoo Kimura and Tornoko Ohta, On some principles governing molecular evolution, Proc. Nat. Acad. Sci, *71*, 2848–2852 (1974).
7 J. L. King and T. H. Jukes, Non-Darwinian evolution, Science *164*, 788–798 (1969).
8 R. F. Doolittle and B. Blombäck, Amino-acid sequence investigations of fibrinopeptides from various mammals: evolutionary implications, Nature *202*, 147–152 (1964).
9 R. F. Doolittle, G. L. Wooding, Y. Lin, and M. Riley, Hominoid evolution as judged by fibrinopeptide structures, J. Mol. Evol. *1*, 74–83 (1971).
10 L. S. B. Leakey, The relationship of African apes, man, and the Old World Monkeys, Proc. Nat. Acad. Sci. *67*, 746–748 (1970).
11 A. C. Wilson and V. M. Sarich, A molecular time scale for human evolution, Proc. Nat. Acad. Sci. *63*, 1088–1093 (1969).
12 M. O. Dayhoff and R. V. Eck, Paleobiochemistry, in *Organic Geochemistry*, G. Eglinton and M. T. J. Murphy, eds. (New York, Heidelberg, and Berlin: Springer-Verlag, 1969), p. 205.
13 D. I. Arnon, Ferredoxin and Photosynthesis, Science *149*, 1460–1470 (1965).
14 L. E. Mortenson, Ferredoxin and ATP, requirements for nitrogen fixation in cell-free extracts of *Clostridium pasteurianum*, Proc. Nat. Acad. Sci. *52*, 272–279 (1964).
15 R. Bachofen, B. B. Buchanan, and D. I. Arnon, Ferredoxin as a reductant in pyruvate synthesis by bacterial extract, Proc. Nat. Acad. Sci. *51*, 690–694 (1964).
16 R. V. Eck and M. O. Dayhoff, Evolution of the structure of ferredoxin based on living relics of primitive amino acid sequences, Science *152*, 363–366 (1966).
17 D. O. Hall, R. Cammack, and K. K. Rao, Role for ferredoxin in the origin of life and biological evolution, Nature *233*, 136–138 (1971).
18 S. M. Siegel, K. Roberts, H. Nathan, and O. Daly, Living relative of the microfossil *Kakabekia*, Science, *156*, 1231–1234 (1967).
19 Barbara Z. Siegel, *Kakabekia*, a review of its physiological and environmental features and their

relation to its possible ancient affinities, in *Chemical Evolution of the Early Precambrian* (New York: Academic Press, 1977), pp. 143–154.

Chapter 11. Two Kinds of Life

1 Cited in E. B. Wilson, *The Cell in Development and Heredity,* 3d ed. (New York: Macmillan, 1925), p. 45.

2 R. Altmann, *Die Elementarorganismen und ihre Beziehungen zu dem Zellen* (Leipzig: Veit und Comp., 1890).

3 Cited in Wilson, *The Cell in Development,* p. 738.

4 I. E. Wallin, The mitochondrion problem, Am. Naturalist *57,* 255–261 (1923).

5 H. Ris and W. Plaut, Ultrastructure of DNA-containing areas in the chloroplast of Chlamydomonas, J. Cell Biol. *13,* 383–391 (1962).

6 P. R. Bell and K. Mühlethaler, Evidence for the presence of deoxy-ribonucleic acid in the organelles of the egg cells of *Pteridium aquilium,* J. Mol. Biol. *8,* 853–862 (1964); D. J. L. Luck and E. Rich, DNA in mitochondria of *Neurospora crassa,* Proc. Nat. Acad. Sci. *52,* 931–938 (1964); M. M. Nass, S. Nass, and B. A. Atzelius, The general occurrence of mitochondria DNA, Exp. Cell Res. *37,* 516–539 (1965); F. L. Schuster, A deoxyribose nucleic acid component in mitochondria of *Didymium nigripes,* a slime mold, Exp. Cell Res. *39,* 329–345 (1965).

7 L. Margulis, *Early Life* (Boston: Science Books International, Inc., 1982), p. 95.

8 P. L. Carpenter, *Microbiology,* 3d. ed. (Philadelphia: W. B. Saunders Co., 1967), p. 15.

9 L. Margulis, *Symbiosis in Cell Evolution* (San Francisco: W. H. Freeman and Co., 1981), p. 327.

10 W. F. Doolittle, The cyanobacterial genome, its expression and the control of that expression, in *Advances in Microbial Physiology, 20,* A. H. Rose and J. G. Morris, eds. (London: Academic Press, 1979), pp. 1–102.

11 Margulis, *Symbiosis in Cell Evolution,* p. 215.

12 P. John and F. R. Whatley, *Paracoccus denitrificans,* a present-day bacterium resembling the hypothetical free-living ancestor of the mitochondrion, Symbiosis Proc. Soc. Expt. Biol. *29,* 39–40 (London: Cambridge University Press, 1975).

13 D. Searcy and R. J. Delange, *Thermoplasma acidophilum* histonelike protein partial amino acid sequence suggestive of homology to eukaryotic histones, Biochim. Biophys. Acta *609,* 197–200 (1980).

14 C. R. Woese, G. E. Fox, L. Zablen, T. Uchida, L. Bonen, K. Pechman, B. J. Lewis, and D. Stahl, Conservation of primary structure in 16S rRNA, Nature *254,* 83–86 (1975).

15 Margulis, *Symbiosis in Cell Evolution,* pp. 285–309.

16 C. E. Clifton, *Introduction to the Bacteria,* 2d ed. (New York: McGraw-Hill, 1958), p. 69.

17 Margulis, *Early Life,* p. 101.

18 Margulis, *Symbiosis in Cell Evolution,* pp. 256–273.

19 P. E. Cloud, Pre-metazoan evolution and the origins of Metazoa, in *Evolution and Environment,* E. T. Drake, ed. (New Haven: Yale University Press, 1968), pp. 1–72.

20 J. R. Nursall, Oxygen as a prerequisite to the origin of the metazoa, Nature *183,* 1170–1172 (1969); L. W. Berkner and L. C. Marshall, The history of oxygenic concentration in the earth's atmosphere, Discussions Faraday Soc. *37,* 122–141 (1964); L. W. Berkner and L. C. Marshall, History of major atmospheric components, Proc. Nat. Acad. Sci. *53,* 1215–1226 (1965); J. S. Levine, Surface solar ultraviolet radiation for paleoatmospheric levels of oxygen and ozone, Origins of Life *10,* 313–323 (1980).

21 Berkner and Marshall, The history of oxygenic concentration; Berkner and Marshall, History of major atmospheric components.

22 L. Margulis, J. C. G. Walker, and M. Rambler, Reassessment of roles of oxygen and ultraviolet light in Precambrian evolution, Nature *264,* 620–624 (1976).

23 C. J. Brock and J. I. Harris, Aspects of the structure and evolution of superoxide dismutases,

in *The Evolution of Metalloenzymes, Metalloproteins and Related Materials,* G. J. Leigh, ed. (London: Symposium Press, 1977), pp. 85–99.

24 Margulis, *Early Life,* p. 75.

25 Margulis, Walker, and Rambler, Reassessment of roles.

Chapter 12. Archaebacteria

1 Robert M. Schwartz and Margaret O. Dayhoff, Origins of prokaryotes, eukaryotes, mitochondria, and chloroplasts, Science *199,* 395–403 (1978); John Barnabas, Robert M. Schwartz, and Margaret O. Dayhoff, Evolution of major metabolic innovations in the Precambrian, Origins of Life *12,* 81–91 (1982).

2 D. W. Emerich and R. H. Burris, Interactions of heterologous nitrogenase components that generate catalytically inactive complexes, Proc. Nat. Acad. Sci. *73,* 4369–4373 (1976).

3 H. D. Peck, Jr., in *Evolution in the Microbial World,* M. J. Carlile and J. J. Skehel, eds. (London: Cambridge University Press, 1974), pp. 241–262.

4 M. Schidlowski, Antiquity and evolutionary status of bacterial sulfate reduction: sulfur isotope evidence, Origins of Life *9,* 299–311 (1979).

5 Barnabas, Schwartz, and Dayhoff, Evolution of major metabolic innovations.

6 V. DeMoulin, Protein and nucleic acid sequence data and phylogeny, Science *205,* 1036–1038 (1979).

7 G. E. Fox, E. Stakebrandt, R. B. Hespell, J. Gibson, J. Maniloff, T. A. Dyer, R. S. Wolfe, W. E. Balch, R. S. Tanner, L. J. Magrum, L. B. Zablen, K. Blakemore, R. Gupta, L. Bonen, B. J. Lewis, D. A. Stahl, K. R. Luehrsen, K. N. Chen, and C. R. Woese, The phylogeny of prokaryotes, Science *209,* 457–463 (1980).

8 C. Woese, M. Sogin, D. Stahl, B. J. Lewis, and L. Bonen, A comparison of the 16S ribosomal RNA's from mesophilic and thermophilic bacilli: some modifications in the Sanger method of RNA sequencing, J. Mol. Evol. *7,* 197–213 (1976); G. E. Fox, K. R. Pechman, and C. R. Woese, Comparative cataloging of 16S ribosomal ribonucleic acid: molecular approach to procaryotic systematics, Int. J. Syst. Bacteriol. *27,* 44–57 (1977); T. Uchida, L. Bonen, H. W. Schaup, B. J. Lewis, L. Zablen, and C. R. Woese, The use of ribonuclease U$_2$ in RNA sequence determination, J. Mol. Evol. *3,* 63–77 (1974).

9 O. Kandler, Zellwandstrukturen bei Methanbakterien zur Evolution der Prokaryoten, Naturwiss. *66,* 95–105 (1979).

10 J. B. Jones, B. Bowers, T. C. Stadtman, *Methanococcus vanielli:* ultrastructure and sensitivity to detergent and antibiotics, J. Bact. *130,* 1404–1406 (1977); R. L. Weiss, Subunit cell wall of *Sulfolbus acidocaldarius,* J. Bact. *118,* 275–284 (1974).

11 O. Kandler and H. König, Chemical composition of the peptidoglycan-free cell walls of methanogenic bacterium, Arch. Microbiol. *118,* 141–152 (1978).

12 G. Darland, T. D. Brock, W. Samsonoff, and S. F. Conti, A thermophilic acidophilic mycoplasma isolated from a coal refuse pile, Science *170,* 1416–1418 (1970).

13 M. Kates, Ether-linked lipids in extremely halophilic bacteria, in *Ether Lipids, Chemistry and Biology,* F. Snyder, ed. (New York: Academic Press, 1972), pp. 351–398; T. A. Langworthy, Long-chain diglycerol tetraethers from thermoplasma acidophilum, Biochim. Biophys. Acta *487,* 37–50 (1977); S. C. Kushwaba, M. Kates, G. D. Sprott, and I. C. P. Smith, Novel complex polar lipids from the methanogenic archaebacterium *Methanospirillum hungeteii,* Science *271,* 1163–1164 (1981).

14 R. Gupta and C. R. Woese, Unusual modification patterns in the transfer RNAs of archaebacteria, Curr. Microbiol. *4,* 245–249 (1980).

15 C. R. Woese, Archaebacteria, Sci. Amer. *244*(6), 98–122 (1981); C. R. Woese, Archaebacteria and cellular origins: an overview, Zbl. Bakt. Hyg., I. Abt. Orig. C *3,* 1–17 (1982).

16 A. T. Matheson, W. Moller, R. Amons, and M. Yaguchi, Comparative studies on the structure of ribosomal proteins with emphasis on the alanine-rich acidic ribosomal A protein, in *Ribosomes:*

Structure, Function, and Genetics, G. Chamblis, G. R. Craven, J. Davies, L. Kahan, and M. Nomura, eds. (Baltimore: University Park Press, 1980), pp. 297–332.

17 W. Zillig, K. O. Stetter, and D. Janekovic, DNA-dependent RNA polymerase from Halobacterium holobium, Europ. J. Biochem. *91*, 193–199 (1978); S. Sturm, U. Schönefeld, W. Zillig, D. Janekovic, and K. O. Stetter, Structure and function of the DNA-dependent RNA polymerase of the archaebacterium *Thermoplasma acidophilum*, Zbl. Bakt. Hyg., I. Abt. Orig. C *1*, 12–25 (1980).

18 C. R. Woese, A proposal concerning the origin of life on the planet earth, J. Mol. Evol. *13*, 95–101 (1979).

Chapter 13. Energetics

1 F. Lipmann, Metabolic generation and utilization of phosphate bond energy, Adv. Enzymol. *1*, 99–162 (1941).

2 Albert L. Lehninger, *Bioenergetics,* 2d ed. (Reading, Mass.: W. A. Benjamin, Inc., 1971), p. 42.

Chapter 14. The Driving Force

1 R. Hill, Oxygen produced by isolated chloroplasts, Proc. Roy. Soc. B *127*, 192–210 (1939).

2 D. I. Arnon, M. B. Allen, and F. R. Whatley, Photosynthesis by isolated chloroplasts, Nature *174*, 394–396 (1954).

3 H. A. Krebs and W. A. Johnson, The role of citric acid in intermediate metabolism in animal tissues, Enzymologia *4*, 148–156 (1937).

Chapter 15. The Question of Genesis

1 *Atlas of Protein Sequence and Structure,* vol. 5, M. O. Dayhoff, ed. (Washington D.C.: National Biomedical Research Foundation, 1972), p. 50.

2 R. J. Britten and D. E. Kohne, Repeated sequences of DNA, Science *161*, 529–540 (1968).

3 *Atlas of Protein Sequence and Structure,* vol. 5, pp. D35–D36.

4 H. J. Morowitz, Biological self-replicating systems, Prog. Theor. Biol. *1*, 35–58 (1967).

5 Wayne F. Frair, Life in a test tube? Quart. J. Creation Res. Soc. *5*, 34–41 (1968).

6 M. Dixon and E. C. Webb, *Enzymes* (New York: Academic Press, 1958), pp. 667–670.

7 H. Fraekel-Conrat and R. C. Williams, Reconstitution of active tobacco mosaic virus from the inactive protein and nucleic acid components, Proc. Nat. Acad. Sci. *41*, 690–698 (1955).

Chapter 16. The Essentials of Life

1 R. W. Holley, J. Aggar, G. A. Everett, J. T. Madison, M. Marquisse, S. H. Merrill, J. R. Penwick, and A. Zamir, Structure of a ribonucleic acid, Science *147*, 1462–1465 (1965).

2 B. G. Barrell and B. F. C. Clark, *Handbook of Nucleic Acid Sequences* (Oxford: Joynson-Bruvvers, 1974).

3 L. E. Orgel, Evolution of the genetic apparatus, J. Mol. Biol. *38*, 381–393 (1968).

4 Sidney W. Fox, The proteinoid theory of the origin of life and competing ideas, Am. Biol. Teacher *36*, no. 3, 161–172 (1974).

5 I. Asimov, *Fact and Fancy* (New York: Avon Books, 1972), pp. 11–20.

6 J. B. McClendon, Elemental abundances as a factor in the origins of mineral nutrient requirements, J. Mol. Evol. *8*, 175–195 (1976).

7 D. O. Hall, R. Cammack, and K. K. Rao, Role for ferredoxin in the origin of life and biological evolution, Nature *233*, 136–138 (1971).

8 McClendon, Elemental abundances.

9 L. Margulis, *Origin of Eukaryotic Cells* (New Haven: Yale University Press, 1970), pp. 7, 8.

10 F. H. C. Crick and L. E. Orgel, Directed panspermia, Icarus *19*, 341–346 (1973); W. R.

Chappell, R. R. Meglen, and D. D. Runnelis, Comments on "Directed panspermia," Icarus *21*, 513–515 (1974); T. H. Jukes, Seawater and the origins of life, Icarus *21*, 516–517 (1974); L. E. Orgel, Reply: "Comments on 'Directed panspermia' " and "Seawater and the origins of life," Icarus *21*, 518 (1974); A. Banin and J. Navrot, Origin of life: clues from relations between chemical composition of living organisms and natural environments, Science *189*, 550–551 (1975).

Chapter 17. The Search for the Building Blocks

1 E. Pflüger, Ueber die physiologische Verbrennung in den lebendigen Organismen, Arch. gesam. Physiol. 10 (1875).

2 A. I. Oparin, *Proiskhozhdenie zhizni (The Origin of Life)* (Moscow: Ixd. Moskovskiy Rabochiy, 1924).

3 A. I. Oparin, *The Origin of Life*, trans. S. Morgulis (New York: Macmillan, 1938).

4 J. B. S. Haldane, The origin of life, Rationalist Annual, 148–153 (1928), repr. in *Science and Human Life* (New York: Harper Brothers, 1933).

5 H. N. Russell, Astrophys. J. *70*, 11 (1929), cited in Virginia Trimble, The origin and abundances of the chemical elements, Rev. mod. Phys. *47*, 877–986 (1975).

6 J. D. Bernal, *The Physical Basis of Life* (London: Routledge and Kegan Paul, 1951).

7 J. W. Williams, Problems in protein chemistry, in Colloid Chemistry, Ann. Rev. Phys. Chem. *2*, 403–424 (1951).

8 F. Sanger and E. O. P. Thompson, The amino-acid sequence in the glycyl chain of insulin, Biochem. J. *53*, 366–374 (1953).

9 W. M. Garrison, D. C. Morrison, J. G. Hamilton, A. A. Benson, and M. Calvin, Reduction of carbon dioxide in aqueous solution by ionizing radiation, Science *114*, 416–418 (1951).

10 S. L. Miller, A production of amino acides under possible primitive earth conditions, Science *117*, 528–529 (1953).

11 J. D. Watson and F. H. C. Crick, A structure for deoxyribose nucleic acid, Nature *171*, 737–738 (1953).

12 S. L. Miller, Production of some organic compounds under possible primitive earth conditions, J. Am. Chem. Soc. *77*, 2351–2361 (1955); S. L. Miller, The mechanism of synthesis of amino acids by electric discharge, Biochim. Biophys. Acta *23*, 480–489 (1957); S. L. Miller, The formation of organic compounds on the primitive earth, Ann. N.Y. Acad. Sci. *69*, 260–275 (1957).

13 T. E. Pavlovskaya and A. G. Pasynskii, The original formation of amino acids under the action of ultraviolet rays and electric discharges, in *The Origin of Life on the Earth*, F. Clark and R. L. M. Synge, eds. (New York: Pergamon Press, 1959), pp. 151–157; A. N. Terenin, Photosynthesis in the shortest ultraviolet, in ibid., pp. 136–139.

14 P. H. Abelson, Paleobiochemistry, Carnegie Inst. of Washington Yearbook, no. 53 (1955–56).

15 W. Groth and H. V. Weyssenhof, Photochemical formation of organic compounds from mixtures of simple gases, Planet. Space Sci. *2*, 79–85 (1960).

16 M. Ya. Dodonova and A. I. Sidorova, Photosynthesis of amino acids from a mixture of simple gases under the action of short-wave ultraviolet radiation, Biophys. *6*, 164–175 (1961).

17 Carl Sagan and Bishun N. Khare, Long-wavelength ultraviolet photoproduction of amino acids on primitive Earth, Science *173*, 417–420 (1971).

18 K. Dose and B. Rajewsky, Strahlenchemische Bildung von Aminen und Aminocarbonsäuren, Biochim. Biophys. Acta *25*, 225–226 (1957).

19 N. Friedmann and S. L. Miller, Phenylalanine and tyrosine synthesis under primitive earth conditions, Science *166*, 766–767 (1969).

20 N. Friedmann, W. J. Haverland, and S. L. Miller, Prebiotic synthesis of the aromatic and other amino acids, in *Molecular Evolution I. Chemical Evolution and the Origin of Life*, R. Buvet and C. Ponnamperuma, eds. (Amsterdam: North-Holland Publ. Co., 1971), pp. 123–135.

21 A. Bar-Nun, N. Bar-Nun, S. H. Bauer, and C. Sagan, Shock synthesis of amino acids in simulated primitive environments, Science *168*, 470–473 (1970).
22 W. W. Rubey, Geologic history of sea water, Bull. Geol. Soc. Am. *62*, 1111–1148 (1951).
23 Philip H. Abelson, Chemical events on the primitive earth, Proc. Nat. Acad. Sci. *55*, 1365–1372 (1966).
24 Ibid.
25 J. Oró and S. S. Kamat, Amino acid synthesis from hydrogen cyanide under possible primitive earth conditions, Nature *190*, 442–443 (1961).
26 C. Ponnamperuma and N. W. Gabel, Current status of chemical studies on the origin of life, Space Life Sci. *1*, 64–96 (1968).
27 Abelson, Chemical events.
28 Cited in A. I. Oparin, *Genesis and Evolutionary Development of Life* (New York: Academic Press, 1968), p. 66.
29 P. Kropokin, in *Sovetskaya Geologiya (Soviet Geology)*, Coll. 47, (Gos. Nauchno-tekhu. Ind., 1955), p. 104.

Chapter 18. Nucleosides, Nucleotides, and ATP

1 J. Oró, Synthesis of adenine from ammonium cyanide, Biochem. Biophys. Res. Comm. *2*, 407–412 (1960).
2 J. Oró and A. P. Kimball, Synthesis of purines under possible earth conditions I. Adenine from hydrogen cyanide, Arch. Biochem. Biophys. *94*, 217–227 (1961).
3 J. Oró and S. S. Kamat, Amino acid synthesis from hydrogen cyanide under possible primitive earth conditions, Nature *190*, 442–443 (1961).
4 C. U. Lowe, R. W. Rees, and R. Markham, Synthesis of complex organic compounds from simple precursors: formation of amino acids, amino acid polymers, fatty acids and purines from ammonium cyanide, Nature *199*, 219–222 (1963).
5 R. A. Sanchez, J. P. Ferris, and L. E. Orgel, Cyanoacetylene in prebiotic synthesis, Science *154*, 784–785 (1966).
6 J. P. Ferris, J. C. Joshi, and J. G. Lawless, Chemical evolution XXIX. Pyrimidines from hydrogen cyanide, Biosystems *9*, 81–86 (1977).
7 Sanchez, Ferris, and Orgel, Cyanoacetylene in prebiotic synthesis; J. P. Ferris, R. A. Sanchez, and L. E. Orgel, Studies in prebiotic synthesis III. Synthesis of pyrimidines from cyanoacetylene and cyanate, J. Mol. Biol. *33*, 693–704 (1968).
8 J. Oró and A. C. Cox, Non-enzymic synthesis of 2-deoxyribose, Fed. Proc. *21*, 80 (1962).
9 C. Ponnamperuma and P. Kirk, Synthesis of deoxyadenosine under simulated primitive earth conditions, Nature *203*, 400–401 (1964).
10 S. L. Miller and L. E. Orgel, *The Origins of Life on the Earth* (Englewood Cliffs, N.J.: Prentice-Hall, Inc., 1974), p. 113.
11 R. A. Sanchez and L. E. Orgel, Studies in prebiotic synthesis V. Synthesis and photoanomerization of pyrimidine nucleoside, J. Mol. Biol. *47*, 531–543 (1970).
12 T. V. Waenheldt and S. W. Fox, Phosphorylation of nucleosides with polyphosphoric acid, Biochim. Biophys. Acta *134*, 1–8 (1967).
13 B. Mason, *Principles of Geochemistry* (New York: John Wiley, 1966), p. 100.
14 S. L. Miller and M. Parris, Synthesis of pyrophosphates under primitive earth conditions, Nature *204*, 1248–1250 (1964).
15 A. W. Schwartz, Phosphate: solubilization and activation on the primitive earth, in *Molecular Evolution*, R. Buvet and C. Ponnamperuma, eds. (Amsterdam: North-Holland Publ. Co., 1971), pp. 207–223.
16 C. Ponnamperuma and R. Mack, Nucleotide synthesis under possible primitive earth conditions, Science *148*, 1221–1223 (1965).

17 R. Lohrmann and L. E. Orgel, Urea-inorganic phosphate mixtures as prebiotic phosphorylating agents, Science *171*, 490–494 (1971).

18 R. Österberg, L. E. Orgel, and R. Lohrmann, Further studies of urea-catalyzed phosphorylation reactions, J. Mol. Evol. *2*, 231–234 (1973).

19 J. Brooks and G. Shaw, *Origin and Development of Living Systems* (New York: Academic Press, 1973).

Chapter 19. Polypeptides

1 F. Hofmeister, Über Bau and Gruppierung der Eiweisskörper, Ergeb. Physiol. *1*, 759–802 (1902).

2 E. Fischer, Ueber einige Derivate des Glykocolls, Alanins und Leucins, Ber. *35*, 1095–1106 (1902).

3 H. Schiff, Ueber Polyaspartsäuren, Ber. *30*, 2449–2459 (1897).

4 L. Balbiano, Ueber ein neues Glykocollanhydrid, Ber. *34*, 1501–1504 (1901); L. Balbiano and D. Trasciatti, Ueber ein neues Derivat des Glykocolls, Ber. *33*, 2323–2326 (1900).

5 T. Curtius and A. Benrath, Ueber benzoyl-pentaglycerlamide Essigsäure (γ-Säure), Ber. *37*, 1279–1310 (1904).

6 W. H. Carothers, Polymers and polyfunctionality, Trans. Faraday Soc. *32*, 39–53 (1936).

7 S. W. Fox and M. Middlebrook, Anhydrocopolymerization of amino acids under the influence of hypothetically primitive terrestrial conditions, Fed. Proc. *13*, 211 (1954).

8 Sidney W. Fox and Milton Winitz, Enzymic synthesis of peptide bonds IV. Effects of variation in substrate structure on relative extents of synthesis of benzoylamine acid anilides as catalyzed by papain and ficin, Arch. Biochem. *35*, 419–427 (1952).

9 S. W. Fox, K. Harada, and A. Vegotsky, Thermal polymerization of amino acids and a theory of biochemical origins, Experientia *15*, 81–84 (1959).

10 S. W. Fox, Evolution of protein molecules and thermal synthesis of biochemical substances, Am Scientist *44*, no. 4, 347–359 (1956).

11 S. W. Fox and K. Harada, Thermal polymerization of amino acids to a product resembling protein, Science *128*, 1214 (1958).

12 S. W. Fox and K. Dose, *Molecular Evolution and the Origin of Life* (San Francisco: W. H. Freeman and Co., 1972).

13 S. W. Fox and K. Harada, The thermal copolymerization of amino acids common to protein, J. Am. Chem. Soc. *82*, 3745–3751 (1960).

14 D. L. Rohlfing and S. W. Fox, Catalytic activities of thermal polyanhydro-α-amino acids, in *Advances in Catalysis 20*, 373–418 (1969).

15 J. Kovacs, I. Koenyves, and A. Pusztai, Darstellung von Polyasparaginsäuren (Polyaspartsäuren aus dem thermischen Autokondensations Produkt der Asparaginsäure), Experientia *9*, 459–460 (1953).

16 Fox and Harada, Thermal copolymerization of amino acids common to protein.

17 J. Kovacs, in *Polyamino Acid, Polypeptides and Proteins*, M. Stahman, ed. (Madison: University of Wisconsin Press, 1962), pp. 53–54.

18 K. A. Grossenbacher and C. A. Knight, Amino acids, peptides and spherules obtained from "primitive earth" gases in a sparking system, in *The Origins of Prebiological Systems*, S. W. Fox, ed. (New York: Academic Press, 1965), pp. 173–186.

19 P. Stiffel, R. D. Minard, and C. N. Matthews, cited in C. N. Matthews, The origin of proteins: heteropolypeptides from hydrogen cyanide and water, Origins of Life *6*, 155–162 (1975).

20 C. Sagan and B. N. Khare, Long-wavelength ultraviolet photoproduction of amino acids on primitive Earth, Science *173*, 417–420 (1971).

21 Clifford N. Matthews and Robert E. Moser, Prebiological protein synthesis, Proc. Nat. Acad. Sci. *56*, 1087–1094 (1966); Clifford N. Matthews and Robert E. Moser, Peptide synthesis from hydrogen cyanide and water, Nature *215*, 1230–1234 (1967).

22 R. E. Moser, A. K. Claggert, and C. N. Matthews, Peptide formation from diaminomalonitrile (HCN tetramer), Tetrahedron Letters *15*, 1599–1603 (1968).

23 C. N. Matthews, the origin of proteins: heteropolypeptides from hydrogen cyanide and water, Origins of Life *6*, 155–162 (1975).

24 I. G. Draganic, Z. D. Draganic, S. Jovanovic, and S. V. Ribnikar, Infrared spectral characterization of peptidic material produced by ionizing radiation in aqueous cyanides, J. Mol. Evol. *10*, 103–109 (1977).

25 R. Minard, W. Yang, P. Varma, J. Nelson, and C. Matthews, Heteropolypeptides from poly-α-cyanoglycine and hydrogen cyanide: a model for the origin of proteins, Science *190*, 387–389 (1975).

26 James P. Ferris, HCN did not condense to give heteropolypeptides on primitive earth, Science *203*, 1135–1137 (1979).

27 C. I. Simionescu, F. Dénes, and I. Negulescu, Abiotic synthesis and the properties of some protobiocopolymers, J. Polymer Sc. Polymer Symp. *64*, 281–304 (1978).

28 C. Krewson and J. Couch, The hydrolysis of nicotinonitrile by ammonia, J. Am. Chem. Soc. *65*, 2256–2257 (1943).

29 S. Akabori, in *Origin of Life on Earth*, A. I. Oparin, ed. (Oxford: Pergamon Press, 1959), pp. 189–196; S. Akabori, Kagaku (Science) *25*, 54 (1955).

30 J. D. Bernal, *The Physical Basis of Life*, (London: Routledge and Kegan Paul, 1951).

31 M. Paecht-Horowitz and A. Katchalsky, Polycondensation of amino acid phosphoanhydrides II. Polymerization of proline adenylate at constant phosphoanhydride concentration, Biochim. Biophys. Acta *140*, 14–23 (1967); M. Paecht-Horowitz, J. Berger, and A. Katchalsky, Prebiotic synthesis of polypeptides by heterogeneous polycondensation of amino acid adenylates, Nature *228*, 636–639 (1970); M. Paecht-Horowitz, The mechanism of clay catalyzed polymerization of amino acid adenylates, Biosystems *9*, 93–98 (1977).

32 A. Katchalsky, Prebiotic synthesis of biopolymers on inorganic templates, Naturwiss. *60*, 215–220 (1973).

33 M. Eigen, Self-organization of matter and the evolution of biological macromolecules, Naturwiss. *58*, 465–523 (1971).

34 K. Dose, Peptides and amino acids in the primordial hydrosphere, Biosystems *6*, 224–228 (1975).

35 N. Lahav and S. Chang, The possible role of solid surface area in condensation reactions during chemical evolution: reevaluation, J. Mol. Evol. *8*, 357–380 (1976).

36 Bernal, *Physical Basis of Life*.

37 H. A. Ireland, in *Silica in Sediments*, H. A. Ireland, ed., Spec. Publs. Soc. Econ. Paleont. Miner., no. 7 (Tulsa, 1959).

38 G. Millot, Géologie des Argiles altérations, sédimentologie géochimie (Paris: Masson et Cie, 1970); in *Geology of Clays*, trans. W. R. Parrand and Helene Paquet (London: Chapman and Hall, 1970), pp. 45–48.

39 Armin Weiss, Replication and evolution in inorganic systems, Angew. Chem. Int. Ed. Engl. *20*, 850–860 (1981).

Chapter 20. The Enzyme Mystique

1 L. Pauling, R. B. Corey, and H. R. Branson, The structure of proteins: two hydrogen-bonded helical configurations of the polypeptide chain, Proc. Nat. Acad, Sci. *37*, 205–211 (1951).

2 J. Jollès, J. Jauregui-Adell, I. Bernier, and P. Jollès, La structure chimique du lysozyme de blanc d'oeuf de poule: Étude détaillée, Biochim. Biophys. Acta *78*, 668–689 (1963).

3 R. E. Canfield and A. K. Liu, The disulfide bonds of egg white lysozyme (Muramidase), J. Biol. Chem. *240*, 1997–2002 (1965).

4 D. C., Phillips, The hen egg-white lysozyme molecule, Proc. Nat. Acad. Sci. *57*, 484–495 (1967).

5 R. E. Dickerson, X-ray analysis and protein structure, in *The Proteins,* 2d ed., H. Neurath, ed. (New York: Academic Press, 1964), vol. 2, p. 634.

6 *Atlas of Protein Sequence and Structure,* vol. 5, M. O. Dayhoff, ed. (Washington, D.C.: National Biomedical Research Foundation, 1972), pp. D133, D134.

7 C. Ponnamperuma and N. W. Gabel, Current status of chemical studies on the origin of life, Space Life Sciences *1,* 64–96 (1968); C. N. Matthews and R. E. Moser, Peptide synthesis from hydrogen cyanide and water, Nature *215,* 1230–1234 (1967); C. U. Lowe, R. W. Rees, and R. Markham, Synthesis of complex organic compounds from simple precursors: formation of amino acids, amino acid polymers, fatty acids and purines from ammonium cyanide, Nature *199,* 219–222 (1963).

8 R. V. Eck and M. O. Dayhoff, Evolution of the structure of ferredoxin based on living relics of primitive amino acid sequences, Science *152,* 363–366 (1966).

Chapter 21. Gene Splinting

1 G. Schramm, H. Grotsch, and W. Pollmann, Non-enzymatic synthesis of polysaccharides, nucleosides and nucleic acids and the origin of self-reproducing systems, Angew. Chem. Intl. Ed. *1,* 1–7 (1962); A. W. Schwartz and S. W. Fox, Condensation of cytidylic acid in the presence of polyphosphoric acid, Biochim. Biophys. Acta *134,* 9–16 (1967).

2 A. Kornberg, *Enzymatic Synthesis of Deoxyribonucleic Acid* (New York: Academic Press, 1961).

3 J. Sulston, R. Lohrmann, L. E. Orgel, and H. T. Miles, Non-enzymatic synthesis of oligoadenylates on a polyuridylic acid template, Proc. Nat. Acad. Sci. *59,* 726–733 (1968).

4 B. J. Weismann, R. Lohrmann, L. E. Orgel, E. Schneider-Bernloehr, and J. E. Sulston, Template-directed synthesis with adenosine-5'-phosphorimidazolide, Science *161,* 387 (1968).

5 M. Renz, R. Lohrmann, and L. E. Orgel, Catalysts for the polymerization of adenosine cyclic 2', 3'-phosphate on a poly (U) template, Biochim. Biophys. Acta *240,* 463–471 (1973).

6 M. S. Verlander, R. Lohrmann, and L. E. Orgel, Catalysts for the self-polymerization of adenosine cyclic 2', 3'-phosphate, J. Mol. Evol. *2,* 303–306 (1973).

7 C. M. Tapiero and J. Nagyvary, Prebiotic formation of cytidine nucleotides, Nature *231,* 42–43 (1971).

8 J. Skoda, J. Morávek, and J. Kopecký, Sixth FEBS Meeting Abstr., 433 (1969).

9 O. Pongs and P. O. P. T'so, Polymerization of 5'-deoxyribonucleotides with β-imidazolyl-4(5)-propanoic acid, Biochem. Biophys. Res. Comm. *36,* 475–481 (1971).

10 J. Ibanez, A. P. Kimball, and J. Oró, the effect of imidazole, cyanamide, and polyornithine on the condensation of nucleotides in aqueous solutions, in *Chemical Evolution and the Origin of Life,* R. Buvet and C. Ponnamperuma, eds. (Amsterdam: North-Holland Publ. Co., 1971), pp. 171–179.

11 Ibid.

12 J. D. Ibanez, A. P. Kimball, and J. Oró, Condensation of mononucleotides by imidazole, J. Mol. Evol. *1,* 112–114 (1971).

13 E. Stephen-Sherwood, D. G. Odom, and J. Oró, The prebiotic synthesis of deoxythymidine oligonucleotides, J. Mol. Evol. *3,* 323–330 (1974).

14 K. L. Agarwal, H. Büch, M. H. Caruthers, N. Gupta, H. G. Khorana, A. Kleppe, A. Kumar, E. Ohtsuka, W. L. Rajbhanary, J. H. Van de Sande, V. Sgaramella, H. Weber, and T. Yamada, Total synthesis of the gene for an alanine transfer ribonucleic acid from yeast, Nature *237,* 27–34 (1970).

Chapter 22. "Particles of Life"

1 M. Traube, Experimente zur Theorie der Zellenbildung und Endosmose, Arch. für Anat. Physiol. und Wiss. Med. *87,* 129–165 (1867).

2 O. Bütschli, *Untersuchungen über microscopische Schäume und das Protoplasma* (Leipzig, 1892).

3 L. Rhumbler, Aus dem Lückengebiet zwischen organismischer und anorganismischer Materie, Ergebn. Anat. Entwicklungsgesch. *15*, 1–38 (1906).

4 S. Le Duc, *The Mechanism of Life*, trans. W. Deane Butcher (New York: Rebman Co., 1914).

5 Martin Kuckuck, *L'univers, être vivant. La solution des problèmes de la matière et la vie à l'aide de la biologie universelle* (Geneva: Librarie Kündig, 1911).

6 Wo. Ostwald and R. Köhler, Ueber die flüssig-flüssige Endmischung von Gelatine durch Sulfosalizylsäure and über die Beziehungen dieses Systems zur Phasenregel, Kolloid Z. *43*, 131–150 (1927).

7 H. G. Bungenberg de Jong, Die Konzervation und ihre Bedeutung für die Biologie, Protoplasma *15*, 110–173 (1932).

8 A. I. Oparin, *The Origin of Life*, trans. S. Morgulis (New York: Macmillan, 1938).

9 A. I. Oparin, Biochemical processes in the simplest structures, in *The Origin of Life on the Earth*, F. Clark and R. L. M. Synge eds. (New York: Pergamon Press, 1959), pp. 428–436.

10 A. I. Oparin, the pathways of the primary development of metabolism and artificial modeling of this development in coacervate drops, in *The Origins of Prebiological Systems*, S. W. Fox, ed. (New York: Academic Press, 1965), pp. 331–348.

11 A. L. Herrara, A new theory of the origin and nature of life, Science *96*, 14 (1942).

12 K. A. Grossenbacher and C. A. Knight, Amino acids, peptides, and spherules obtained from "primitive earth" gases in a sparking system, in *The Origins of Prebiological Systems*, pp. 173–186.

13 C. I. Simionescu, F. Dénes, and M. Macoveau, Synthesis of some amino acids, sugars and peptides in cold plasma. Electron-microscopic studies on some proteid forms (III), Biopolymers *12*, 237–241 (1973).

14 A. E. Smith, J. J. Silver, and G. Steinman, Cell-like structures from simple molecules under simulated primitive earth conditions, Experientia *24*, 36–38 (1968).

15 S. W. Fox, K. Harada, and J. Kendrick, Production of spherules from synthetic proteinoid and hot water, Science *129*, 1221–1223 (1959).

16 S. W. Fox and S. Yuyama, Effects of the Gram stain on microspheres from thermal polyamino acids, J. Bacteriol. *85*, 279–283 (1963).

17 S. W. Fox, R. J. McCauley, and A. Wood, A model of primitive heterotrophic proliferation, Comp. Biochem. Physiol. *20*, 773–778 (1967).

18 S. W. Fox and S. Yuyama, Abiotic product of primitive protein and formed microparticles, Ann. N.Y. Acad. Sci. *108*, 487–494 (1963).

19 L. L. Hsu, S. Brooke, and S. W. Fox, Conjugation of proteinoid microspheres: a model of primordial communication, Curr. Mod. Biol. *4*, 12 (1971).

20 L. L. Hsu, and S. W. Fox, Interaction between diverse proteinoids and microspheres in simulation of primordial evolution. Biosystems *8*, 89–101 (1977).

21 J. D. Bernal, *The Origin of Life* (London: Weidenfeld and Nicholson, 1967), p. 125.

22 S. W. Fox, R. J. McCauley, P. Mongomery, T. Fukushima, K. Harada, and C. R. Windsor, Membrane-like properties in microsystems assembled from synthetic protein-like polymers, in *Physical Principles of Biological Membranes*, F. Snell, J. Wolken, G. Iverson, and J. Lam, eds. (New York: Gordon and Breach, 1969), pp. 417–432.

23 W. Stillwell, Facilitated diffusion of amino acids across bimolecular lipid membranes as a model for selective accumulation of amino acids in a primordial protocell, Biosystems *8*, 111–117 (1976).

24 Lynn Margulis, ed., *Origins of Life*, (New York: Gordon and Breach, 1970), p. 157.

25 S. W. Fox, Simulated natural experiments in spontaneous organization of morphological units from proteinoid, in *The Origins of Prebiological Systems*, pp. 361–382.

26 K. Bahadur and S. Ranganayaki, Synthesis of Jeewanu. The units capable of growth, multipli-

cation and metabolic activity. I. Preparation of units capable of growth and division and having metabolic activity, Zentbl. Bakt. Parasitkde. (II) *117*, 567–574 (1964).

27 K. Bahadur, Synthesis of Jeewanu. The units capable of growth, multiplication and metabolic activity. III. Preparation of microspheres capable of growth and division by building and having metabolic activity with peptides prepared thermally, Zentbl. Bakt. Parasitkde. (II) *117*, 585–602 (1962).

28 A. I. Oparin, *Genesis and Evolutionary Development of Life* (New York: Academic Press, 1968), pp. 127–151.

29 S. W. Fox, Origin of the cell: experiments and premises, Naturwiss. *60*, 359–368 (1973).

Chapter 23. The Vital Envelope

1 A. D. Bangham and R. W. Horne, Negative staining of phospholipids and their structural modification by surface active agents as observed in the electron microscope, J. Mol. Biol. *8*, 660–668 (1964).

2 W. Stillwell, Facilitated diffusion of amino acids across bimolecular lipid membranes as a model for selective accumulation of amino acids in a primordial protocell, Biosystems *8*, 111–117 (1976).

3 William Stillwell and Aruna Rau, Primordial transport of sugars and amino acids via Schiff bases, Origins of Life *11*, 243–254 (1981).

4 William Stillwell, Facilitated diffusion as a method for selective accumulation of materials from the primordial oceans by a lipid-vesicle protocell, Origins of Life *10*, 277–292 (1980).

5 R. J. Goldacre, Surface films, their collapse on compression, the shapes and sizes of cells and the origin of life, in *Surface Phenomena in Chemistry and Biology*, J. F. Danielli, K. G. A. Pankhurst, and A. C. Riddiford, eds. (New York: Pergamon Press, 1958), pp. 278–298.

6 W. R. Hargreaves, S. J. Mulvhill, and D. W. Deamer, Synthesis of phospholipids and membranes in prebiotic conditions, Nature *266*, 78–80 (1977).

7 W. R. Hargreaves and D. W. Deamer, Origin and early evolution of bilayer membranes, in *Light Transducing Membranes: Structure, Function, and Evolution*, D. W. Deamer, ed. (New York: Academic Press, 1978).

8 D. W. Deamer and G. L. Burchfield, Encapsulation of macromolecules by lipid vesicles under simulated prebiotic conditions, J. Mol. Evol. *18*, 203–204 (1982).

9 D. W. Deamer, personal communication.

Chapter 24. The Emergence of Cells

1 D. A. Usher, Early chemical evolution of nucleic acids: a theoretical model, Science *196*, 311–313 (1977).

2 W. Stillwell, On the origin of photophosphorylation, J. Theor. Biol. *65*, 479–497 (1977).

3 *Atlas of Protein Sequence and Structure*, vol. 5, M. O. Dayhoff, ed. (Washington, D.C.: National Biomedical Research Foundation, 1972), pp. 111–117.

4 F. H. C. Crick, S. Brenner, A. Klug, and G. Pieczenik, A speculation on the origin of protein synthesis, Origins of Life *7*, 389–397 (1976).

5 R. V. Eck and M. O. Dayhoff, Evolution of the structure of ferredoxin based on living relics of primitive amino acid sequences, Science *152*, 363–366 (1966).

6 P. J. McLaughlin and M. O. Dayhoff, Eukaryotes versus prokaryotes: an estimate of evolutionary distance, Science *168*, 1469–1471 (1970).

7 *Atlas of Protein Sequence and Structure*, vol. 5, p. 116.

Chapter 25. The Phenomenal Cell

1 Philip H. Abelson, Chemical events on the primitive earth, Proc. Nat. Acad. Sci. *55*, 1365–1372 (1966).

2 H. J. Morowitz, Physical background of cycles in biological systems, J. Theor. Biol. *13*, 60–62 (1966).

3 Ch. Degani and M. Halmann, Chemical evolution of carbohydrate metabolism, Nature *216*, 1207 (1967).

4 K. Decker, K. Jungermann, and R. K. Thauer, Wege der Energiegewinnung in Anaerobiern, Angew. Chem. *82*, 153–173 (1970).

5 R. Bachofen, B. B. Buchanan, and D. I. Arnon, Ferredoxin as a reductant in pyruvate synthesis by a bacterial extract, Proc. Nat. Acad. Sci. *51*, 690–694 (1964).

6 W. E. Balch, L. Magrum, G. Fox, R. Wolfe, and C. Woese, An ancient divergence among bacteria, J. Mol. Evol. *9*, 305–311 (1977).

7 Carl R. Woese, A comment on methanogenic bacteria and the primitive ecology, J. Mol. Evol. *9*, 369–371 (1977).

8 F. R. Japp, Stereochemistry and vitalism, Nature *58*, 452 (1898).

9 J. B. Van't Hoff, *The Arrangement of Atoms in Space*, 2d ed. (London: Longmans, 1898).

10 W. Kuhn and E. Braun, Photochemische Erzeugung optisch aktivar Stoffe, Naturwiss. *17*, 227–228 (1929).

11 K. Harada, Origin and development of optical activity of organic compounds on primordial earth, Naturwiss. *57*, 114–119 (1970).

12 F. Vester, T. I. V. Ulbricht, and H. Krauch, Optische Aktivität und die Paritätsverletzung im β-Zerfall, Naturwiss. *46*, 68 (1957).

13 M. Goldhaber, L. Grodzins, and A. W. Sunyar, Evidence for circular polarization of Bremsstrahlung produced by beta rays, Phys. Rev. *106*, 826–829 (1957).

14 H. P. Noyes, W. A. Bonner, and J. A. Tomlin, On the origin of biological chirality via natural beta-decay, Origins of Life *8*, 21–23 (1977).

15 S. L. Miller and L. E. Orgel, *The Origins of Life on the Earth* (Englewood Cliffs, N.J.: Prentice-Hall, Inc., 1974), p. 171.

16 G. Wald, Origin of optical activity, Ann. N.Y. Acad. Sci. *69*, 352–368 (1957).

17 Harold F. Blum, *Time's Arrow and Evolution*, 2d ed. (New York: Harper and Bros., 1962), p. 172.

Chapter 26. Other Ways, Other Places

1 M. E. Jones and F. Lipmann, Chemical and enzymatic synthesis of carbamyl phosphate, Proc. Nat. Acad. Sci. *46*, 1194–1205 (1960).

2 I. S. Kulaev, Inorganic polyphosphates in evolution of phosphorus metabolism, in *Chemical Evolution and the Origin of Life*, R. Buvet and C. Ponnamperuma, eds. (Amsterdam: North-Holland Publ. Co., 1971), pp. 458–465.

3 H. Baltscheffsky, Inorganic pyrophosphate and the origin and evolution of biological energy transformation, in *Chemical Evolution and the Origin of Life*, pp. 466–473.

4 *Atlas of Protein Sequence and Structure*, vol. 5, M. O. Dayhoff, ed. (Washington, D.C.: National Biomedical Research Foundation, 1972), p. 117.

5 W. Gevers, H. Kleinkauf, and F. Lipmann, The activation of amino acids for biosynthesis of gramicidin S, Proc. Nat. Acad. Sci. *60*, 269–276 (1968); Gevers, Kleinkauf, and Lipmann, Interrelationship between activation and polymerization of gramicidin S biosynthesis, Proc. Nat. Acad. Sci. *62*, 226–233 (1969).

6 R. Roskoski, Jr., H. Kleinkauf, W. Gevers, and F. Lipmann, Nonribosomal polypeptide synthesis: activation and condensation of amino acids in tyrocidin formation, Fed. Proc. *29*, 468 (1970).

7 W. Gevers, H. Kleinkauf, and F. Lipmann, Peptidyl transfers in gramicidin S biosynthesis from enzyme-bound thioester intermediates, Proc. Nat. Acad. Sci. *63*, 1335–1342 (1969).

8 R. A. Gibbons and G. D. Hunter, Nature of the scrapie agent, Nature *215*, 1041–1043 (1967).

9 J. S. Griffith, Self-replication and scrapie, Nature *215*, 1043–1044 (1967).

10 J. B. McClendon, Elemental abundances as a factor in the origins of mineral nutrient requirements, J. Mol. Evol. *8*, 175–195 (1976).
11 A. G. Cairns-Smith, The origin of life and the nature of the primitive gene, J. Theoret. Biol. *10*, 53–88 (1965).
12 P. J. E. Peebles, The structure and composition of Jupiter and Saturn, Astrophys. J. *140*, 328–347 (1964).
13 Cited in R. Wildt, H. J. Smith, E. E. Saltpeter, and A. G. W. Cameron, The planet Jupiter, Physics Today *16*, 19–23 (May 1963).
14 Ibid.
15 F. Woeller and C. Ponnamperuma, Organic synthesis in a simulated Jovian atmosphere, Icarus *10*, 386–392 (1969).

Chapter 27. Organic Compounds in the Universe

1 J. J. Berzelius, Ueber Meteorsteine, Ann. Phys. und Chem. *33*, 113 (1834).
2 F. Wöhler, Über die Bestandteile des Meteorsteines von Kaba in Ungarn, Sitz. Math. Naturwiss. Akad. Wien *33*, 205–209 (1859).
3 S. Cloëz, Note sur la composition chimique de la pierre météorique d'Orgueil, Compt. rend. *58*, 986 (1864); S. Cloëz, Analyse chimique de la pierre météorique d'Orgueil, Compt. rend. *59*, 37 (1864).
4 P. Berthelot, La matière carbonneuse de la météorite d'Orgueil purifiée autant que possible par les dissolvants, s'est ensuite oxydée entierement, J. parkt. Chem. *106*, 254 (1869).
5 G. Mueller, The properties and theory of genesis of the carbonaceous complex within the cold bokevelt meteorite, Geochim. Cosmochim. Acta *4*, 1–10 (1953).
6 B. Mason, Origin of chondrules and chondritic meteorites, Nature *186*, 230–231 (1960).
7 B. Nagy, W. G. Meinschein, and D. J. Hennessy, Mass spectroscopic analysis of the Orgueil meteorite: evidence for biogenic hydrocarbons, Ann. N.Y. Acad. Sci. *93*, 25–35 (1961).
8 G. Claus and B. Nagy, A microbiological examination of some carbonaceous chondrites, Nature *192*, 594–596 (1963).
9 Edward Anders, Meteoritic hydrocarbons and extraterrestrial life, Ann. N.Y. Acad. Sci. *93*, 649–664 (1963).
10 M. H. Briggs and G. Mamikunian, Organic constituents of carbonaceous chondrites, Space Sci. Rev. *1*, 647–682 (1963).
11 M. H. Studier, R. Hayatsu, and E. Anders, Organic compounds in carbonaceous chondrites, Science *149*, 1455–1459 (1965).
12 E. Anders, Origin, age, and composition of meteorites, Space Sci. Rev. *3*, 583–714 (164).
13 Studier, Hayatsu, and Anders, Organic compounds in carbonaceous chondrites.
14 M. O. Dayhoff, E. R. Lippincott, and R. V. Eck. Thermodynamic equilibria in prebiological atmospheres, Science *146*, 1461–1464 (1964).
15 A. C. Cheung, D. M. Rank, C. H. Townes, D. D. Thornton, and W. J. Welch, Detection of NH₃ molecules in the interstellar medium by their microwave emission, Phys. Rev. Letters *21*, 1701–1705 (1968).
16 A. C. Cheung, D. M. Rank, C. H. Townes, D. D. Thornton, and W. J. Welch, Detection of water in interstellar regions by its microwave radiation, Nature *221*, 626–628 (1969).
17 L. E. Snyder and D. Buhl, Water-vapor clouds in the interstellar medium, Astrophys. J. *155*, 165–170 (1969).
18 R. W. Wilson, K. B. Jeffreys, and A. A. Penzias, Carbon monoxide in the Orion Nebula, Astrophys. J. *161*, 143–144 (1970).
19 L. E. Snyder and D. Buhl, Radio emission from HCN, IAU Circular no. 2251 (1970).
20 L. E. Snyder, D. Buhl, B. Zuckerman, and P. Palmer, Microwave detection of interstellar formaldehyde, Phys. Rev. Letters *22*, 679–681 (1969).
21 B. B. Turner, Radio emission from interstellar cyanoethylene, IAU Circular no. 2268 (1970).

22 David Buhl, Chemical constituents of interstellar clouds, Nature *234*, 332–334 (1971).

23 F. Fitch, H. P. Scharcz, and E. Anders, "Organized elements" in carbonaceous chondrites, Nature *193*, 1123–1125 (1962).

24 Michael H. Briggs and G. Barrie Kitto, Complex organic microstructures in the Mokoia meteorite, Nature *193*, 1126–1127 (1962).

25 B. Nagy, G. Claus, and D. Hennessy, Organic particles embedded in minerals in the Orgueil and Ivuna carbonaceous chondrites, Nature *193*, 1129–1133 (1962).

26 F. W. Fitch and E. Anders, Observations on the nature of the "organized elements" in carbonaceous chondrites, Ann. N.Y. Acad. Sci. *108*, 495–513 (1963).

27 E. T. Degens and M. Bajor, Amino acids and sugars in the Bruderheim and Murray meteorites, Naturwiss. *49*, 605–606 (1962).

28 I. R. Kaplan, E. T. Degens, and J. H. Reuter, Organic compounds in stony meteorites, Geochim. Cosmochim. Acta *27*, 805–834 (1963).

29 Paul B. Hamilton, Amino acids in hands, Nature *205*, 284–285 (1965).

30 J. Oró and H. B. Skewes, Free amino acids on human fingers: the question of contamination in microanalysis, Nature *207*, 1042–1045 (1965).

31 K. Kvenvolden, J. Lawless, K. Pering, E. Peterson, J. Flores, C. Ponnamp.ruma, I. R. Kaplan, and C. Moore, Evidence for extraterrestrial amino acids and hydrocarbons in the Murchison meteorite, Nature *228*, 923–926 (1970).

32 K. A. Kvenvolden, J. G. Lawless, and C. Ponnamperuma, Non-protein amino acids in the Murchison meteorite, Proc. Nat. Acad. Sci. *68*, 486 (1971).

32 G. Dungworth, Optical configuration and the racemization of amino acids in sediments and fossils, Chem. geol. *17*(2), 135–153 (1976).

34 S. W. Fox and K. Harada, Accumulated analyses of amino acid precursors in returned lunar samples, in *Proceedings of the Fourth Lunar Science Conference*, Cosmochim. Acta, vol. 2, suppl. 4 (New York: Pergamon Press, 1973), pp. 2241–2248.

Chapter 28. Gaia

1 Preston Cloud and Aharon Gibor, The oxygen cycle, Sci. Amer. *223*(3), 110–123 (1970).

2 J. E. Lovelock, Gaia as seen through the atmosphere, Atmos. Environ. *6*, 579–580 (1972); J. E. Lovelock and L. Margulis, Homeostatic tendencies of the Earth's atmosphere, Origins of Life *1*, 12–22 (1974); James Lovelock and Lynn Margulis, Atmosphere homeostasis: The Gaia Hypothesis, Tellus *26*, 1–10 (1974); James Lovelock and Sydney Epton, The quest for Gaia, The New Scientist *65*, 304–309 (1975); Lynn Margulis and James E. Lovelock, The atmosphere as circulatory system of the biosphere—The Gaia Hypothesis, The CoEvolution Quarterly (Summer, 1975), pp. 31–41.

Glossary

activation energy
: An energy barrier to a spontaneous reaction.

active site
: Portion of an enzyme engaged in catalytic activity.

aerobes
: Organisms that derive energy by oxidative phosphorylation.

anaerobes
: Organisms that derive their energy solely from fermentation.

andesitic rocks
: Plutonic rocks intermediate between granite and basalt in silica content.

archaebacteria
: Group of procaryotes thought to have diverged very early from the lines that led to respiring heterotrophic bacteria.

Archean era
: Division of geologic history between the Hadean and Proterozoic eras 3.8 to 2.6 billion years ago.

basalt
: Rock formed when magma solidifies at or near the surface.

Cambrian period
: Geologic age extending from 570 to 500 million years ago.

carbon dioxide fixation
: Reduction of atmospheric carbon dioxide.

Cenozoic era
: Division of geologic history extending from the end of the Mesozoic era 65 million years ago to the present.

chitinozoan
: Unicellular heterotroph.

chloroplast
: Eucaryotic organelle containing the photosynthesis apparatus.

chloroxybacteria	Oxygen-releasing photosynthetic pro-caryote lacking the phycobiliproteins of the cyanobacteria.
coacervate	Colloidal particle formed from natural polymers.
codon	Three-nucleotide sequence on the DNA molecule that translates to an amino acid.
coelenterate	Organism at the tissue level of development and lacking organs.
eucaryote	Organism with gene-bearing structures in chromosomes contained in a membrane-bound nucleus.
fermentation	Enzymic breakdown of organic matter in the absence of molecular oxygen.
gene	Segment of DNA molecule that translates to a protein.
genetic code	The correlation of amino acids to the codons from which they are translated.
genome	The combined gene and chromosome complement.
gneiss	Laminated metamorphic rock corresponding in composition to granite or other plutonic rock.
granite	Plutonic rock having 66 percent or more silica.
greenhouse effect	Heating due to infrared radiation that cannot penetrate a covering that is transparent to other radiation.
greenstone belts	Foundation rocks formed by the rapid solidification of lava under water.
Hadean era	Earliest age of the earth extending from the earth's formation to the appearance of a stable crust 3.8 billion years ago.
heterotroph	Organism that obtains nourishment from organic matter.
igneous rocks	Rocks formed by the crystallization of magma.
island arcs	Volcanic islands formed by extrusion of magma from subducted crustal material.
kerogen	Intractable and insoluble organic residue.

liposome	Spherical particle formed by the enclosure of bimolecular lipid layer.
"living fossils"	Existing organisms that have not undergone appreciable morphological change over geologic time.
magma	Molten rock between the mantle and the crust.
mantle	Layer of the earth that extends from the rim of the core to the base of the crust.
Mesozoic era	Division of geologic history between the Paleozoic and Cenozoic eras representing the age of dinosaurs and marine and flying reptiles.
metazoan	Any of a group comprising all animals except the protozoans.
mitochondrion	Eucaryotic organelle containing the enzymic mechanism for oxidative phosphorylation.
molecular evolution	Mutational change of homologous molecules.
ninhydrin	Reagent for the detection of free amino and carboxyl groups in proteins, peptides, and amino acids giving a blue color under the proper conditions.
nitrogen fixation	Reduction of atmospheric nitrogen.
optical activity	The rotation of polarized light.
organelle	A membrane-bound structure inside a eucaryotic cell.
orogeny	The process of mountain building.
oxidative phosphorylation	The formation of ATP by the oxidation of pyruvate to carbon dioxide and water with molecular oxygen.
ozone	Triatomic oxygen produced continuously in the outer layers of the atmosphere by the action of solar ultraviolet radiation on the oxygen in the air.
Paleozoic era	Division of geologic history between the Proterozoic and the Mesozoic eras and extending from the earliest fossils to and including land plants, amphibians, and the earliest reptiles.
Pasteur point	Concentration level of free oxygen intolerant to obligate anaerobes.

phylum   A basic anatomical type.

plutonic rocks   Rocks formed by the slow solidifica-
                                tion of magma beneath the surface.

procaryote   An organism without a membrane-
                                bound nucleus.

Proterozoic era   Division of geologic history from the
                                Archean to the Paleozoic eras, 2.6 bil-
                                lion to 570 million years ago.

solar wind   Stream of subatomic particles emanat-
                                ing from the sun.

stromatolite   Limestone, dolomite, or siliceous
                                structure of biogenic origin formed in
                                intertidal waters.

vesicle   Enclosed bimolecular lipid particle of
                                biological origin.

Select Bibliography

Bernal, J. D. *The Origin of Life*. London: Weidenfeld and Nicholson, 1967.

Blum, H. F. *Time's Arrow and Evolution*. New York: Harper and Row, 1962.

Brooks, J., and G. Shaw. *Origin and Development of Living System*. New York: Academic Press, 1973.

Calvin, M. *Chemical Evolution*. New York and Oxford: Oxford University Press, 1969.

Fox, S. W., ed. *The Origins of Prebiological Systems*. New York: Academic Press, 1965.

Fox, S. W., and K. Dose. *Molecular Evolution and the Origins of Life*. San Francisco: W. H. Freeman & Co., 1972.

Kenyon, D. H., and G. Steinman. *Biochemical Predestination*. New York: McGraw-Hill Co., 1969.

Kvenvolden, K. A., ed. *Geochemistry and the Origin of Life*. New York: Dowden, Hutchinson, and Ross, Inc., 1974.

Miller, S. L., and L. Orgel. *The Origins of Life on the Earth*. Englewood Cliffs, N.J.: Prentice-Hall, Inc., 1974.

Oparin, A. I. *Genesis and Evolutionary Development of Life*. New York: Academic Press, 1968.

———. *The Chemical Origin of Life*. Springfield, Ill.: Chas. C. Thomas, 1964.

Rutten, M. G. *The Origin of Life*. Amsterdam: Elsevier Publishing Co., 1971.

Journals on the Origin of Life

Origins of Life
Biosystems
J. Molecular Evolution
J. Precambrian Research

Author Index

The italicized numbers refer to the pages where the reference article is cited.

Subject Index

actinobacteria, 47
activation energy, 110, 111
adenine, abiotic synthesis of, 160
adenosine 2′, 3′-phosphate, 191
adenosine triphosphate. *See* ATP
aerobic breakdown of glucose, 126–28
amino acids: chemical structures, 179; from ammonium cyanide, 160; in fingerprints, 252; in meteorites, 251–53; polymerization on clay, 174, 175; precursors in moon rocks, 254; produced in the Miller experiment, 153; substitution rate in proteins from mutation, 75; D-amino acids, naturally occurring, 233
ammonium cyanide: as precursor of amino acids and purines, 160
anaerobes: earliest life forms, 29; oxygen tolerance level, 31
anaerobic breakdown of glucose, 123, 124; energy yield, 128
archaebacteria: cell wall composition, 106; difference in tRNAs, 106; membrane composition, 106; occurrences of, 107, 108
Archean era: atmosphere composition, 12; temperature, 14
Astasia, 90
Athiorhodacea: nutritional requirement, 26; resemblance to earliest organisms, 26
atmosphere: composition of early, 9; during Archean era, 12; on primordial earth, 156; when life began, 28
ATP: biological energetics, 112–15; chemical structure, 164; energy of hydrolysis, 112; prebiotic occurrence, 165; production by chloroplasts, 121; three means of synthesis, 117–28

bacteria, size and composition of, 87, 134
banded iron formation: last major episode, 19, 30; origin, 19
Beck Spring Dolomite fossils, 24
Beckspringia, 23; biological cell, minimal requirements, 147, 148; models for, 198–205; smallest autonomous, 135
biological systems: basic structural units, 207; commonalities, 131
Biot, Jean Baptiste, 230
Bitter Spring Formation, microfossils in, 20
blue-green algae. *See* cyanobacteria
Büchner, Eduard: discovers the enzymic basis of fermentation, 54
Bulawayan stromatolites, date of formation of, 20
Brown, Robert: discovers the nucleus of the cell, 40

calcium sulfate, oldest deposits of, 31
Cambrian period: definition, 15; fossils, 15, 16; temperature, 36; thickness of fossil strata, 17
carbodiinides, as condensing agents, 189, 194
carbon dioxide: concentration in early atmosphere, 10; greenhouse effect due to, 10
carbon dioxide fixation: earliest occurrence, 23; photosynthesis and, 125
carbonaceous chondrites, organic matter in, 246–54
cells: as basis of life, 40, 41; early models of, 198–205; number in the human body, 48
Chaos chaos, number of mitochondria in, 44
Charnia, as Proterozoic fossils, 16
Chase, Martha, 60
chitinozoans, earliest fossils of, 24
Chlamydomonas, coding for proteins in chloroplasts of, 91
Chlorobium: nitrogen fixation by, 102; resemblance to earliest organisms, 26
chlorophyll: chemical structure, 119; in photosynthesis, 118
chloroplasts: as symbionts, 90, 91; ATP synthesis by, 121; DNA in, 89; electron micrograph, 122; number in plants, 45
chloroxybacteria, 47
chordate, earliest fossils of, 36